TRANSPORTATION INFRASTRUCTURE - ROADS, HIGHWAYS, BRIDGES, AIRPORTS AND MASS TRANSITSERIES

HIGHWAYS

FEDERAL-STATE PARTNERSHIPS AND PRACTICES

TRANSPORTATION INFRASTRUCTURE - ROADS, HIGHWAYS, BRIDGES, AIRPORTS AND MASS TRANSITSERIES

Additional books in this series can be found on Nova's website under the Series tab.

Additional E-books in this series can be found on Nova's website under the E-book tab.

TRANSPORTATION INFRASTRUCTURE - ROADS, HIGHWAYS, BRIDGES, AIRPORTS AND MASS TRANSITSERIES

HIGHWAYS

FEDERAL-STATE PARTNERSHIPS AND PRACTICES

BARBARA E. ALLEN
AND
JASON G. GRAY
EDITORS

nova publishers
New York

For permission to use material from this book please contact us:
Telephone 631-231-7269; Fax 631-231-8175
Web Site: http://www.novapublishers.com

Library of Congress Cataloging-in-Publication Data

ISBN: 978-1-62257-733-0

Published by Nova Science Publishers, Inc. † New York

CONTENTS

PREFACE

The U.S. Department of Transportation (DOT) provides about $40 billion to the states annually to build and maintain highways and bridges through the federal-aid highway program. This book describes the process and factors that could affect highway project time frames; examines state DOTs' views on the benefits and challenges of the provisions to expedite highway projects established in SAFETEA-LU and describes additional initiatives that state DOTs and FHWA have implemented to expedite the completion of highway projects.

Chapter 1 – The U.S. Department of Transportation (DOT) provides about $40 billion to the states annually to build and maintain highways and bridges through the federal-aid highway program. While this program has grown and changed over time, the federal-state relationship has been consistently one of "partnership" since 1916. DOT's FHWA has offices in all 50 states that have developed close working relationships with states. Legislation approved by the Senate in March 2012 would establish a more performance-based highway program, introducing performance measures for highways and bridges and requiring FHWA to monitor states' progress in meeting those measures. As requested, GAO examined (1) how the federal-aid highway program and FHWA's oversight have changed over time; (2) the extent to which FHWA's partnership approach produces benefits; (3) the extent to which FHWA's partnership approach poses risks; and (4) how FHWA's partnership with state DOTs could affect a transition toward a performance-based highway program. To do this work, GAO conducted site visits and a survey, reviewed relevant documentation, and interviewed FHWA and state officials.

Chapter 2 – Projects to construct, improve, and repair roads and bridges are fundamental to meeting the nation's mobility needs. However, completing highway projects—which generally involves four phases consisting of (1) planning, (2) preliminary design and environmental review, (3) final design and right-of-way acquisition, and (4) construction—can sometimes take a long time. In 2005, SAFETEA-LU established provisions to help expedite highway projects, including streamlining some portions of the environmental review process, allowing states to assume greater environmental review responsibilities under certain conditions, and establishing efforts that permitted delegation of some authority from the federal government to states. GAO was asked to (1) describe the process and factors that could affect highway project time frames, (2) examine state DOTs' views on the benefits and challenges of the provisions to expedite highway projects established in SAFETEA-LU, and (3) describe additional initiatives that state DOTs and FHWA have implemented to expedite the completion of highway projects. GAO surveyed officials from 52 state DOTs, including all states, the District of Columbia, and Puerto Rico; interviewed officials at FHWA, state DOTs, and federal resource agencies (agencies tasked with protecting natural, historic, or cultural resources); and analyzed legislation, regulations, and other reports and publications. U.S. DOT provided technical comments on a draft of this report, which GAO incorporated as appropriate.

In: Highways
Editors: B. E. Allen and J. G. Gray

ISBN: 978-1-62257-733-0

Chapter 1

HIGHWAY INFRASTRUCTURE: FEDERAL-STATE PARTNERSHIP PRODUCES BENEFITS AND POSES OVERSIGHT RISKS*

United States Government Accountability Office

WHY GAO DID THIS STUDY

The U.S. Department of Transportation (DOT) provides about $40 billion to the states annually to build and maintain highways and bridges through the federal-aid highway program. While this program has grown and changed over time, the federal-state relationship has been consistently one of "partnership" since 1916. DOT's FHWA has offices in all 50 states that have developed close working relationships with states. Legislation approved by the Senate in March 2012 would establish a more performance-based highway program, introducing performance measures for highways and bridges and requiring FHWA to monitor states' progress in meeting those measures. As requested, GAO examined (1) how the federal-aid highway program and FHWA's oversight have changed over time; (2) the extent to which FHWA's partnership approach produces benefits; (3) the extent to which FHWA's partnership approach poses risks; and (4) how FHWA's partnership with state

* This is an edited, reformatted and augmented version of the Highlights of GAO-12-474, a report to the Ranking Member, Subcommittee on Highways and Transit, Committee on Transportation and Infrastructure, House of Representatives, dated April 2012.

DOTs could affect a transition toward a performance-based highway program. To do this work, GAO conducted site visits and a survey, reviewed relevant documentation, and interviewed FHWA and state officials.

What GAO Recommends

Congress should consider restructuring federal surface transportation programs. Based on GAO's review, there may be areas where national interests are less evident and where opportunities exist to narrow FHWA's responsibilities. Also, DOT should address the risks posed by its partnership approach. DOT generally agreed with the recommendation.

What GAO Found

Over the years, the federal-aid highway program has expanded to encompass broader goals, more responsibilities, and a variety of approaches. As the program grew more complex, the Federal Highway Administration's (FHWA) oversight role also expanded, while its resources have not kept pace. As GAO has reported, this growth occurred without a well-defined overall vision of evident national interests and the federal role in achieving them. GAO has recommended Congress consider restructuring federal surface transportation programs, and for this and other reasons, funding surface transportation remains on GAO's high-risk list.

FHWA benefits from using recognized partnership practices to advance the federal-aid highway program and conduct program oversight—such as clear delineation of roles and responsibilities between FHWA and its state partners and formal and informal conflict resolution—that are recognized as leading practices. FHWA's partnership approach allows it to proactively identify issues before they become problems, achieve cost savings, and gain states' commitment to improve their processes.

FHWA's partnership approach also poses risks. We observed cases where FHWA was lax in its oversight or reluctant to take corrective action to bring states back into compliance with federal requirements, potentially resulting in improper or ineffective use of federal funds. For example, while FHWA has made it a national priority to recoup funds from inactive highway projects—projects that have not expended funds for over 1 year—FHWA officials in three states we visited were reluctant to do so because of concerns about harming their partnership with the state. In other cases, FHWA has shown a lack of independence in decisions, putting its partners' interests above federal interests. For example, FHWA allowed two states to retain unused emergency relief allocations to fund new emergencies, despite FHWA's policy that these funds are made available to other states with potentially higher-priority emergencies. In some instances, FHWA became actively and closely involved in implementing solutions to state problems—this can create a conflict when FHWA later must approve or review the effectiveness of those solutions.

If proposals for a performance-based highway program are adopted, FHWA would have to work with states to develop measures and take corrective action if states do not meet them. FHWA's partnership could help states develop measures, but it would need to mitigate the risks posed by its partnership to ensure corrective action was effective when needed. The fundamental reexamination of surface transportation programs, including the highway program, that GAO previously recommended presents an opportunity to narrow FHWA's responsibilities so that it is better equipped to transition to a performance-based system. GAO identified areas where national interests may be less evident but where FHWA expends considerable time and resources—areas where devolving responsibilities to the states may be appropriate.

ABBREVIATIONS

DBE	Disadvantaged Business Enterprise
DOT	Department of Transportation
FHWA	Federal Highway Administration
MAP-21	Moving Ahead for Progress in the 21st Century
MPO	Metropolitan Planning Organization
NRT	National Review Team
Recovery Act	American Recovery and Reinvestment Act of 2009
TMA	Transportation Management Area

April 26, 2012

The Honorable Peter A. DeFazio
Ranking Member

Subcommittee on Highways and Transit
Committee on Transportation and Infrastructure
House of Representatives

Dear Mr. DeFazio:

Federal funding for highways is provided to the states mostly through a series of formula grant programs collectively known as the federal-aid highway program. Over the years, in response to changing transportation, environmental, and societal goals, the highway program has expanded to encompass broader goals, more responsibilities, and a variety of grant structures. As the highway program grew more complex, the federal government's responsibility to oversee the program expanded as well. The role of the Department of Transportation's (DOT) Federal Highway Administration (FHWA) has grown from overseeing the construction of highway projects to a wide range of activities including overseeing states' transportation planning processes; assessing environmental impacts; and monitoring states' compliance with property acquisition, contracting, civil rights, and other requirements.[1] As we have reported, FHWA has faced challenges in ensuring that federal funds are efficiently and effectively used because the highway program is one in which there is limited federal control—it is a state-administered, federally assisted program.[2] Funds are largely apportioned by formula, and the states enjoy broad flexibility in deciding which projects are supported. Furthermore, for nearly half of federal-aid highway funds, FHWA's responsibility to oversee the design and construction of projects has been assumed by the states.

To adapt to its evolving role, FHWA has relied on its historical partnership with the states in which FHWA and the states work collaboratively to construct highway infrastructure. Specifically, FHWA division offices, which are located in every state, the District of Columbia, and Puerto Rico, have developed close working relationships with state departments of transportation (state DOTs). (For a complete list of division offices, see appendix I.) We have reported that this partnership approach offers benefits, such as more efficient project delivery, but can also create challenges. When an overseer becomes an active partner in a project, the independent perspective important to effective project oversight can be lost.[3]

Legislation approved by the Senate in March 2012[4] would establish performance measures for highways and require FHWA to monitor states' progress in meeting these measures. As a consequence, FHWA's oversight approach could undergo significant change, as FHWA would need to work with the states to develop measurable goals to improve the condition and performance of the nation's highway system and take action when performance measures are not met. Given its history, FHWA is likely to use its partnership with the states to implement such a system. Also, administering a performance-based system would be an addition to FHWA's other duties.

You asked us to examine how FHWA's partnership with state DOTs affects its oversight of the highway program and, in particular, how it might affect the transition to a more performance-based highway program. This report addresses (1) the basis for FHWA's partnership and oversight role in relation to the states, and how the federal-aid highway program and FHWA's oversight have changed over time; (2) the extent to which FHWA's partnership approach produces benefits; (3) the extent to which FHWA's partnership approach poses risks; and (4) how FHWA's partnership with state DOTs could affect a transition toward a performance-based highway program.

To achieve these objectives, we reviewed and analyzed relevant laws and regulations, historical documentation, and stewardship agreements between FHWA and state DOTs that delineate oversight responsibilities. We conducted site visits to 9 of FHWA's 52 division offices, which are located in all 50 states, the District of Columbia, and Puerto Rico, chosen based on a range of criteria, including the size of the state's transportation program and proportion of federal funding, the rural and urban composition of the state, and geographic distribution. During our site visits, we interviewed FHWA division office leadership as well as state officials, when possible, to understand how they characterize their relationship. To obtain a broader perspective than site visits provide, we designed and conducted a survey of the division administrators from each of the 52 FHWA division offices; our response rate was 100 percent. We also conducted 4 discussion groups that included representatives from 38 of 52 state DOTs. Through a literature review, we identified best practices and risks associated with partnering, and used these criteria to assess FHWA's current oversight approach by reviewing information from our audit work. We also analyzed available results from states' 2010 Single Audit findings to independently identify areas of concern in oversight activities. See appendix II for more details on our scope and methodology.

We conducted this performance audit from April 2011 to April 2012 in accordance with generally accepted government auditing standards. Those standards require that we plan and perform the audit to obtain sufficient, appropriate evidence to provide a reasonable basis for our findings and conclusions based on our audit objectives. We believe that the evidence obtained provides a reasonable basis for our findings and conclusions based on our audit objectives.

BACKGROUND

Federal assistance for highway and bridge infrastructure—about $40 billion each year—is distributed through multiple formula and discretionary grant programs collectively known as the federal-aid highway program. The federal-aid highway program is financed through the Highway Trust Fund, a dedicated source of federal revenue based on the "user-pay principle"—that is, users of transportation systems pay for the systems' construction through the federal tax on motor fuels, tires, and trucks.

FHWA uses a decentralized organizational structure to administer the federal-aid highway program, meaning that decision-making authority is largely delegated to FHWA's 52 division offices. FHWA division offices have 10 to 61 staff each, depending on the size of the state's highway program. While there are variations in division office organizational structures, each typically has teams that cover areas such as planning, environment, engineering, technical services, finance, and civil rights. As of February 2012, FHWA had 2,960 staff—1,962 in the field and 998 at headquarters.

FHWA's responsibilities for the federal-aid highway program fall into two broad categories: (1) *advancing the program* and (2) *ensuring compliance with federal law and regulations*. To advance the program, FHWA engages in a range of activities to encourage the effective and efficient use of federal-aid highway funding and assist states in progressing projects through construction to improve the highway system. To accomplish these tasks, FHWA works with states to identify issues, develop and advocate solutions, approve and obligate project funding for eligible activities, and provide technical assistance and training to state DOTs.[5] To ensure that states comply with federal laws and regulations, FHWA, through its division offices, conducts oversight of federally funded projects and reviews state DOT capacity and systems used to administer approved projects. Actual project-level oversight is divided or shared between FHWA and the state. FHWA oversees major interstate highway projects. FHWA division offices and states jointly decide how to divide oversight responsibility for other National Highway System projects.[6] States assume oversight responsibility for projects that are not on the National

Highway System. These can include locally administered projects, which are projects in which a state DOT has given approval to a local public agency (e.g. a city or county) to administer a project or phase of a project such design, property acquisition, or construction.

For those projects where both FHWA and the state make decisions about oversight responsibilities, the respective responsibilities are generally mapped out in a "stewardship agreement." This agreement defines which projects will receive "full" oversight, in which FHWA oversees most aspects of the construction process, or projects in which states assume oversight responsibility (we refer to these as "delegated" projects). Figure 1 describes aspects of oversight that are led by FHWA or the state depending on the status of the project.

To evaluate state DOTs' systems and capacity to administer approved projects, FHWA division offices assess internal controls and processes across programmatic areas such as construction, finance, property acquisition, and locally administered projects. A common tool for this type of oversight is a "process review," which involves an analysis of key program components and processes employed by the state DOT. Typically, this includes a file review of a sample of projects, interviews with relevant state DOT staff, and field reviews when applicable. In addition, FHWA, in conjunction with the Federal Transit Administration, performs a federal certification review of Metropolitan Planning Organizations (MPO), which are responsible for transportation planning in urban areas with populations larger than 50,000[7] every 4 years,[8] meaning that periodically FHWA determines if the organization complies with applicable federal requirements. FHWA division offices have a range of corrective actions they can use if a state does not comply with federal requirements. Among other things, it may withhold funding from all or part of a project, deobligate inactive funds, withhold approval until an issue is resolved, or require corrective action plans.

FHWA's Partnership Role Was Established Almost a Century Ago, but over Time Its Responsibilities Have Expanded

Over the years, the federal-aid highway program has grown to encompass broader goals, more responsibilities, and a variety of approaches; however, the concept of a federal-state partnership has been an integral feature of the highway program since it was established by the Federal Aid Road Act of 1916.[9] This and other early highway legislation established federal-state responsibilities, wherein states select the placement of roads, construct, and maintain them, and the federal government sets standards and provides a portion of the funding. The Federal-Aid Highway Act of 1973 further refined the federal-state relationship by stating that "the authorization of the appropriation of Federal funds…shall in no way infringe on the sovereign rights of the States to determine which projects shall be federally financed" and defined the federal-aid highway program as a "federally assisted State program."[10] This language helped shape FHWA's interpretation of the federal-state relationship, leading to an understanding of its role that anchors FHWA's approach to oversight in partnership and is cited today in FHWA's policy documents.[11] Almost all the division administrators we surveyed described their work with states as a partnership and in ways that emphasized the importance of partnership in carrying out FHWA's mission to advance the transportation program. In addition, as we previously reported, both FHWA and state officials believe that over the years the partnership has helped to build trust and respect between state transportation agencies and FHWA, ensuring that as partners they can accomplish tasks such as planning and building projects more efficiently and effectively.[12]

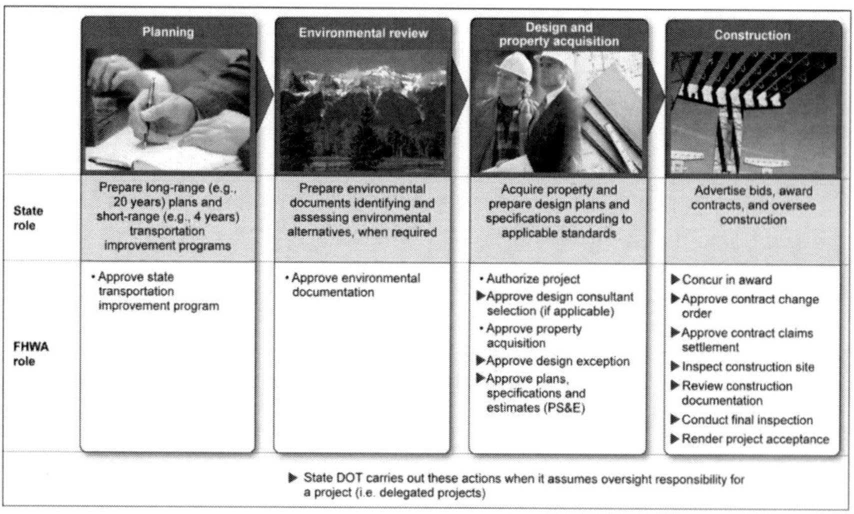

Source: GAO analysis of FHWA information; Art Explosion and Photodisc (photographs).

Figure 1. FHWA and State Oversight and Approval Activities for Full and Delegated Projects.

The goals and scope of the federal-aid highway program expanded during much of the 20th century, as did the roles and responsibilities of FHWA. Initially, the highway program was administered by the Department of Agriculture through the Bureau of Public Roads, a predecessor to FHWA.[13] The bureau focused oversight at the project level to ensure that materials and construction methods met federal standards. The bureau also brought engineering expertise to the states, many of which either lacked skilled engineers or were not ensuring that federal dollars were being used to produce quality construction. The Defense Highway Act of 1941 extended the 1916 act to fund a strategic network of highways, including secondary and feeder routes. Eligibility was extended again in 1944 to include an array of other secondary roads, including rural farm-to-market roads, rural mail and bus routes, county roads, and others that became eligible for post-war federal aid.[14] In 1950, Congress made additional roads—including county, township, and urban roads—eligible for aid.[15] By then, however, the focus of the highway program was turning increasingly to constructing the Interstate Highway System. Designated as mandated in 1944,[16] construction of the system began in earnest with passage of the Federal-Aid Highway Act of 1956[17] and establishment of the Highway Trust Fund, a dedicated funding source deriving revenue primarily from taxes on motor fuels, tires, and trucks to finance the construction of the Interstate system.[18]

Construction of the Interstate remained the focus of the federal-aid highway program in the years that followed, and Congress continued to expand the types of projects eligible for federal funds. In the 1970s, Congress expanded the federal role in bridge infrastructure by making highway bridges located on public roads and longer than 20 feet eligible for federal funds.[19] Congress also expanded the eligibility of federal aid beyond initial construction. Under the 1916 Act and Interstate authorizations, the federal government was to fund the *construction* of highways while *maintenance* was the states' responsibility. However, as the Interstate began to age, Congress allowed states to use federal funds for road maintenance on Interstate highways and all eligible bridges by redefining certain activities—such as resurfacing, rehabilitation, and reconstruction—as capital investments rather than maintenance.[20] By 1991, as a result of changes over the years, use of highway funds was authorized to fund a wide range of transportation enhancement activities, including activities connected with highway beautification, historic preservation, and the establishment of bicycle and pedestrian trails.[21]

In addition to expanding the types of projects eligible for federal highway funds, over time Congress adopted legislation to achieve social goals such as advancing civil rights and environmental protection, and enhancing urban planning and economic development, which affected the federal-aid highway program and FHWA's role and responsibilities. For example, the National Environmental Policy Act of 1969[22] required grant recipients to comply with federal environmental requirements by conducting environmental reviews for federally funded transportation projects. The Federal-Aid Highway Act of 1962 established urban transportation planning as a matter of national interest and required all construction projects to be part of a continuing, comprehensive, and cooperative planning process.[23] Other federal requirements have included requiring compliance with prevailing wage standards applicable to federal contracts under the Davis-Bacon Act, a Disadvantaged Business Enterprise program to enhance participation of women- and minority-owned businesses, and Buy America provisions for acquiring steel and other materials. As the goals of the highway program expanded, FHWA added expertise in its division offices beyond civil engineers and hired economists, right-of-way specialists, planners, historians, ecologists, safety experts, civil rights experts, and others.

In the early 1990s, Congress revised FHWA's role and responsibilities. With the Interstate Highway System nearly complete, Congress created the National Highway System and subsequently states began to assume authority from FHWA. In addition, FHWA's staff was reduced from 3,556 in 1998 to 2,960 in February 2012.in 1998 to 2,960 in February 2012.[24] Recognizing its changing roles, responsibilities, and decline in staff levels, FHWA continued to adapt its oversight approach. In 2006, it began adopting a risk management approach to its oversight, recognizing in part that, while its role had expanded, its resources had not.[25]

FHWA has adapted to changes in demands for its oversight, but its role and responsibilities are complicated by the fact that the current federal approach to surface transportation in general—and to highways in particular—is not working well.[26] The expansion of the program did not result from a specific rationale or plan, but rather an agglomeration of policies and programs since the 1950s without a well-defined overall vision of the national interest and federal role in our surface transportation system.[27] Federal goals and programs are now numerous and sometimes conflicting, and federal roles are unclear. Furthermore, although DOT and FHWA establish national goals and priorities, federal highway funding is apportioned to states without regard to the accomplishment of specific outcomes or the performance of grantees. This makes it difficult to assess the extent to which funding is achieving transportation goals. For these and other reasons, funding surface transportation remains on GAO's high-risk list.[28]

FHWA Benefits from Using Recognized Partnership Practices

Partnering Activities and Practices

In the face of its evolving roles and responsibilities, FHWA has relied on its historical partnership with the states in which FHWA and the states work collaboratively to construct highway infrastructure. FHWA uses partnering activities and practices with the states that are, based on our review and synthesis of partnering literature, recognized as best practices. These activities and practices enable parties to achieve individual and mutually beneficial goals and results, such as expedited project time frames and cost savings. We observed the following examples of successful partnerships:

- *Open and regular communication* includes clear and candid discussions among partners as well as an understanding of the inner workings and decision-making processes of participating organizations. FHWA division and state DOT officials reported having regular formal and informal meetings (at leadership and working levels) as well as frequent contact by e-mail and telephone.
- *Clear delineation of roles and responsibilities* involves understanding individual partner roles as well as articulating responsibilities for joint actions and tasks. FHWA stewardship agreements describe the roles of FHWA divisions and state DOTs. Some stewardship agreements that we examined include detailed matrices addressing factors such as work activities and their frequency, legal authority, and specific division office and state DOT responsibilities.
- *Proactive issue identification and resolution*, in a mutually agreeable way, is closely linked to open and regular communication between partnering members. FHWA officials in several division offices told us that they work closely with their state DOT counterparts to identify

problems early and develop solutions. For example, one division administrator explained that, at times, the division office is forced to tell the state DOT that, because of its approach, a certain portion, or an entire construction project, is not eligible for federal funding. However, the administrator stated that the next question the division staff asks is, "How can we do this?" to work with the state DOT to bring the project into compliance with federal requirements, and therefore allowing the state DOT to use federal funds.

- *Conflict resolution* processes include formal (documented protocols or escalation procedures) and informal (verbal agreements between parties) procedures for how to handle disagreements. FHWA division and state officials discussed their commitment to collaborative problem solving and using informal issue escalation procedures, for example, by elevating problematic issues to the leadership level for resolution. One FHWA division incorporated conflict resolution protocols into its formal partnering agreement with the state DOT. The agreement advocated using face-to-face communication for conflict resolution and outlined procedures for escalating issues.

Partnership Benefits in Advancing the Federal-Aid Highway Program

In our survey and interviews, FHWA division administrators reported that FHWA uses its partnering relationship with state DOTs to advance the federal-aid highway program by ensuring that projects move to construction in a timely fashion, facilitating knowledge transfer, and promoting federal transportation priorities. Specifically, 51 of 52 survey respondents stated that their partnering relationship with their state was very or somewhat important to their ability to achieve the mission of the federal-aid highway program. Most division administrators (44 of 52) also indicated that the partnering relationship produces multiple benefits. Some of these benefits—such as expedited project time frames and cost savings—were noted as positive outcomes of partnership in the literature we reviewed. For example, FHWA officials in a northern state with a short summer construction season told us they work closely with state DOT officials to make timely decisions and move projects along to ensure that construction can be completed during the warmer months. FHWA officials in this state generally conducted non-construction-related process reviews in the winter season so that they and state DOT officials could focus on construction inspections and construction-related process reviews during the warmer months. Similarly, in another state, FHWA and state officials told us that by working together to resolve issues expediently, they were able to complete the environmental review and approval process for a large-scale project to reduce congestion on an important regional highway in about half of the time normally required.

Forty-four of 52 division administrators indicated in response to our survey that partnering was very helpful in facilitating the transfer of technical knowledge. In our site visits, FHWA officials explained that partnering helps FHWA to use technical knowledge transfer to advance the federal-aid highway program by assisting states in addressing technical or programmatic concerns, closing skill gaps, enhancing compliance, and informing decision making. FHWA division officials in one state told us that they had financed a trip by state DOT officials to learn about an alternative interchange design. As a result, the state DOT was able to make interchange improvements without removing and replacing an existing bridge, reducing costs from estimated $10 to $15 million to $3 million. Similarly, when FHWA officials observed crumbling materials used for retaining walls and supporting structures, they brought in the technical expertise of the FHWA Resource Center, which resulted in the state DOT revising its materials specifications to ensure higher-quality materials are used.

Forty-nine of 52 division administrators also indicated in response to our survey that partnering helps the agency to advance federal transportation priorities. One respondent stated that "partnering helps us advance more federal priorities and achieve greater public benefit than simply being parochial authoritarians that refuse to discuss anything that doesn't directly involve a federal dollar or regulation." For example, officials in one FHWA division office developed a business case for an approach to address congestion in a key section of highway as an alternative to the state's planned solution. FHWA promoted its alternative to the state DOT and other stakeholders, and ultimately the state accepted FHWA's approach because the data showed it would be more effective and less costly. In another example, FHWA division officials believed they influenced the state DOT to improve safety by using higher-quality barriers and rumble strips on highways to alert drivers straying off the road. Likewise, to improve safety, officials in another FHWA division office promoted cable barriers on highway medians as a risk-based, lower-cost alternative to concrete barriers.

Similarly, during our site visits, some FHWA officials said that their partnering relationship creates an opportunity to promote projects of national or regional significance within a state. For example, FHWA officials in one division office persuaded the state to address congestion around a toll plaza on a major interstate route. The state did not consider the project a high priority, because it did not affect most state residents as much as out-of-state drivers traveling through the state. However, FHWA division officials were able to persuade the state to construct the project when funding from the American Recovery and Reinvestment Act of 2009 (Recovery Act)[29] became available. One of the FHWA officials commented, "Without partnerships, you lose opportunities to do things that would be good for the taxpayer."

Partnership Benefits in Oversight and Effects on Corrective Action Decisions

FHWA relies on its partnering approach with state DOTs to facilitate oversight of the federal-aid highway program by engaging states in open dialogue about risks and obtaining their buy-in on program improvements. In our survey, 50 of 52 FHWA division administrators said that their division office's partnering relationship was very or somewhat helpful in producing more effective oversight. Officials we spoke with during our site visits offered several illustrations of how partnership improves oversight. One FHWA division administrator told us that the state DOT proactively brings problems to FHWA's attention rather than waiting for FHWA to discover them. We observed that this open dialogue about risks allows FHWA to address issues in a timely fashion and adopt a more responsive and problem-solving attitude. According to one FHWA division official, FHWA worked collaboratively with the state DOT to determine which process reviews to conduct during the year and then conducted the majority of those reviews jointly. According to the FHWA official, this practice strengthened oversight by helping to gain the state DOT's buy-in and commitment to improving its processes, facilitating honest communication about risk areas, and creating an opportunity for FHWA to provide on-the-spot training when problems were identified. In another example, an FHWA division office holds annual meetings with state DOT officials where the two parties determine which projects should receive full oversight and which should be assumed by the state. We observed that this approach can strengthen oversight by allowing FHWA to incorporate the state's perception of risk and weaknesses into their oversight plan by engaging in open dialogue with state officials about risks. Additionally, for each summer construction season, the two parties identify one of the state's regions as a focus for full oversight. This allows each region and their project managers to receive training while their projects, primarily related to pavement preservation, are being reviewed by FHWA.

FHWA's partnership with state DOTs also affects its use of corrective action. FHWA emphasizes working with state DOTs to bring them back into compliance through less stringent corrective action instead of more punitive action. Responses to our survey of division administrators showed that the most frequently used corrective action in the last 3 fiscal years was withholding approval of a particular request until an issue was resolved. According to one division administrator, withholding approval provides the greatest ability to address and resolve a particular issue and encourages the state to take corrective action in a timely manner. Additionally, 43 of 52 division administrators reported that in the last 3 fiscal years they have used the *threat* of a corrective action, which helps to achieve compliance without actual punitive actions. Reportedly, the threat of a corrective action is effective because it communicates the consequences of not complying with federal requirements and helps to bring about problem resolution. In addition, to address deficiencies, 51 of 52 division administrators reported that they had required state DOTs to develop a corrective action plan to outline how the state would change a process or program to comply with federal requirements.

When FHWA moves toward a more punitive corrective action, it is most likely to withhold funding from a part of a project. Withholding partial funding often amounts to not paying for a line item in a project's budget. For example, one division administrator explained that when the state purchases *proprietary* equipment, such as certain types of light posts or signs, when a less expensive *nonproprietary* option is available, FHWA withholds funds for the purchase. The federal-aid highway program is a reimbursement program. As a consequence, if FHWA withholds funds, state DOTs must replace federal funding with state funding. All 52 division offices indicated they withheld partial federal funding from a project in the last 3 fiscal years, and withholding partial funding was the second most frequently used tool for corrective action.

FHWA division offices reported that they rarely use their most punitive corrective actions, such as withholding funding for an entire project or organization. Although 30 division administrators we surveyed reported that they had withheld federal funding from an *entire* project during the past 3 fiscal years, none listed this action as one of their three most frequently used corrective actions. According to FHWA officials, such action is damaging to the state's federal-aid highway program and provokes tension with state DOTs. FHWA officials stated that they see this measure as a last resort and try to use their partnership with the states to elicit compliance. Furthermore, division offices periodically review MPOs, which can receive federal-aid highway funding and implement construction projects. If the office declines to certify an MPO, federal funds for that organization are withheld until the deficiencies identified are corrected.

Navigating the “Gray Areas”

FHWA division officials use partnering practices, such as open and regular communication, with state DOT officials as they exercise administrative discretion[30] in situations where the rules and how to apply them are not clear—situations we refer to as “gray areas.” In administering the federal-aid highway program, FHWA often has discretion to take a less stringent action even when the law permits a harsher one, if circumstances warrant. Such an approach is embodied in a stewardship agreement from one division office, which states that the division office “will make use of available regulatory flexibility when in the public interest.” FHWA officials spend time and effort addressing gray areas, as they seek to make a decision that is not only consistent with federal regulations but also appropriate to the particular facts and circumstances of the situation. For example, the federal regulations governing federal-aid contracts call for state DOTs to use reasonable judgment in evaluating contractors’ good faith effort to hire women- and minority-owned businesses but do not specify the type of documentation contractors must submit to demonstrate their effort.[31] Because there are no specifications on the type of documentation demonstrating a good faith effort, FHWA and state DOT officials must work through this gray area to determine how best to demonstrate their efforts.

In our interviews and observations, we noted that FHWA officials rely on partnering practices with states when federal regulations and FHWA policies leave room for interpretation and discretion, creating gray areas for FHWA officials to resolve. For example, routine roadway maintenance is not eligible for federal reimbursement,[32] but preventive maintenance can be.[33] According to an FHWA division administrator in one state, there is room for interpretation and discretion between the two types of maintenance. This division administrator told us that in recent years the state DOT has sought reimbursement from FHWA for roadway maintenance activities that are typically ineligible for reimbursement—a situation he attributed to the economic environment affecting state budgets. This required time and effort by both FHWA and the state DOT to work through their different interpretations of the regulation. Ultimately, FHWA and the state agreed to develop asset management systems to identify and prioritize preventive maintenance needs in a systematic way. The division administrator explained that this approach would show and document how the maintenance strategies would extend the roadway life and prevent deterioration and higher maintenance costs later, which would make these costs eligible for federal funding.

We also observed FHWA using partnering practices to negotiate the gray areas that may arise when the rules are clear, but practical considerations complicate implementing the rule. When implementing a rule, public officials may need to consider cost-benefit implications, time frames, local economic conditions, or other local circumstances that are not necessarily dealt with explicitly in rules or regulations. For example, FHWA officials in one division office explained that FHWA's regulations require roadside guardrails on National Highway System routes to be a minimum height of 27 ¾ inches from the top of the guardrail to the top of the pavement. The height of a guardrail governs its effectiveness. However, as states overlay pavement with new asphalt to address road deterioration, the height of the guardrail relative to the road surface decreases and the guardrail becomes less effective. This creates practical trade-offs with regard to the costs of guardrail replacement and safety, raising questions regarding whether to use funds to improve the pavement condition of, for example, 20 miles of road without replacing guardrails or to pave fewer miles of road but replace the guardrails to ensure they are at the full height prescribed in the regulation. Both approaches offer safety benefits. Determining the best course of action requires navigating a gray area and requires FHWA to understand the state's priorities, weigh the safety outcomes, and use its partnership with the state to agree on an approach that meets transportation needs and federal responsibilities.

States' Perspectives on Partnering and Oversight

While FHWA officials largely viewed their partnering relationship with state DOTs in positive terms, state officials offered a more tempered response. Specifically, our interviews and discussion groups with officials from 38 state DOTs revealed that while states acknowledged having good working relationships with their FHWA division counterparts, they also expressed some frustrations. On the positive side, state DOT officials appreciated regular and ongoing communication with FHWA officials and characterized FHWA staff as accessible, responsive, and solution-oriented. State DOT officials told us that stewardship agreements were helpful in clarifying roles and expectations and that they consider risk-based oversight to be a strength of the FHWA-state relationship. Officials also appreciated FHWA's help in navigating federal requirements and sharing technical expertise and industry best practices. Furthermore, state officials appreciated that FHWA officials recognized the unique needs, context, and features of their state. State DOT officials participating in our discussion groups asserted that they do not want a "one size fits all" FHWA.

However, state DOT officials' positive feedback about FHWA was tempered with other perspectives on partnering and FHWA decision making. We noted three themes among the comments of state DOT officials when voicing perspectives different from FHWA.

1. *State officials viewed partnership less favorably than FHWA*. Many state DOT officials characterized FHWA's role as providing oversight and enforcing regulations rather than acting as a partner. Some officials indicated FHWA began emphasizing enforcement over partnership around the time of the completion of the Interstate Highway System, as FHWA responded to legislative changes, adopting what many state officials viewed as an audit-focused approach to oversight. According to some state DOT officials, the sense of camaraderie between state DOT and FHWA officials that existed during the building of the Interstate is no longer there, and currently there is "less partnership and more regulation." Reflecting the states' more tempered perspective, one state DOT official characterized the relationship with FHWA as "a partnership within an arranged marriage."
2. *State officials viewed FHWA as imposing personal preferences*. Many state DOT officials told us that FHWA officials routinely imposed personal preferences—for example, questioning particular design solutions—and would threaten to withhold federal funds or approval even though, in the states' view, the approach the state had developed complied with federal standards and regulations. State DOT officials pointed out that these preferences are not covered in the regulations but rather involve professional judgment regarding such factors as cost, appearance, and durability that are not prescribed in regulation and are often unique to a particular construction project. For example,

according to officials at one state DOT, FHWA had a preference for sequencing construction activities in a particular way rather than leaving the decision up to the state DOT, and on this particular project FHWA made its preference a requirement. Some states also said that FHWA gives too much focus to smaller issues and is overly involved in routine matters.

3. *State officials were frustrated by inconsistencies in FHWA's decision making across states.* Many state DOT officials expressed frustration about the inconsistencies they perceived in FHWA's decisions across states. Specifically, several noted that FHWA division offices in other states had been more permissive of certain solutions or requirements compared to the FHWA division office in their state and stated that FHWA does not always use the maximum flexibility it has at its disposal in interpreting federal rules. For example, a state DOT official told us that the division office in his state did not approve a certain material for markings on the state's highways, but he learned that the same material had been approved in 17 other states.

The inconsistencies experienced by state DOTs may not be unreasonable and could stem from the decentralized nature of the federal-aid highway program and the fundamental challenge FHWA and the states face in navigating gray areas on complex projects with unique political, financial, engineering, and other challenges. These complexities likely contribute to FHWA and state DOTs' differing perspectives. For example, we previously reported a case in which a state DOT planned to construct new soundwalls on an existing highway. FHWA noted that the state was planning to widen the road a few years later and that the walls would likely have to be destroyed and rebuilt. FHWA recommended that the state construct the walls at the location envisioned for the widening project, but state DOT officials resisted because of the additional costs to acquire property. FHWA then informed the state that it would only fund construction of the walls once—either at the location along the existing highway as the state had planned or at the location needed once the road was widened.[34] This example illustrates a case in which state officials may have viewed FHWA as imposing its personal preferences and may have been aware of similar situations in other states in which FHWA officials made different decisions inconsistent with this approach. FHWA, on the other hand, may have viewed its decision as exercising professional judgment to promote the most long-term cost-effective solution, consistent with its role as a steward of federal funds.

FHWA'S PARTNERSHIP WITH STATES POSES RISKS IN FHWA'S OVERSIGHT

While successful partnering relationships offer benefits, they also present potential risks, according to the literature we reviewed. First, one partner may grow lax in holding the other to standards. Second, one partner can lose independence in its decisions. We observed cases where FHWA was lax in its oversight by trusting but not verifying state activities and cases where FHWA demonstrated reluctance to take corrective action to bring states back into compliance, which can result in ineffective, wasteful, and potentially improper use of federal funds. We also observed instances in which FHWA sometimes showed a lack of independence in decisions, putting the states' interests above federal ones, and other instances in which FHWA took extraordinary measures to advance the program to the point of becoming actively and closely involved in implementing solutions to state problems. This can create an inherent conflict when FHWA later must review and approve those actions or review their effectiveness. Despite the risks partnership poses, FHWA has good oversight practices in several areas of the federal-aid highway program.

Partnership Created Challenges for FHWA in the Past

We have expressed concerns about the risks posed by FHWA's partnership approach in the past. The Central Artery/Tunnel project in Boston, Massachusetts, provides examples of both lax oversight and a lack of independence that resulted in ineffective and inefficient use of federal funds and damaged FHWA's credibility.[35] This highway project—one of the largest, most complex, and expensive ever undertaken—experienced widely reported cost increases, growing from around \$2.3 billion in the mid-1980s to almost \$15 billion in 2004. From 1995 through 1997, we reported concerns about cost growth and funding gaps on the project and weaknesses in FHWA's efforts to address them and to hold the state accountable.[36] In March 2000, an FHWA task force charged with reviewing FHWA's oversight of the project concluded that "FHWA's long history of strong Federal/State partnerships failed" and that FHWA "had failed to maintain an independent enough relationship with the state to adequately fulfill its oversight role." The task force attributed lax oversight to FHWA placing too much trust in the state, reporting that FHWA's partnership approach failed to achieve independent and critical oversight of the project.[37]

As this example illustrates, although FHWA has experienced partnership risks to its programs in the recent past, FHWA division administrators generally do not recognize the risks of partnering as significant. In our survey, more than half (29) stated they did not believe that partnering creates any risks to their oversight of the federal-aid highway program. Of the remainder, 5 said it may create "some risk," 17 said there was a "slight risk," and only one stated partnering was a "significant risk" to oversight.

FHWA Sometimes Was Lax in Its Oversight or Reluctant to Take Corrective Action

In some instances, FHWA was lax in its oversight in that it did not verify compliance with the requirements of the federal-aid highway program, instead trusting states to ensure its actions were in compliance, which could have resulted in ineffective, wasteful, and potentially improper use of federal funds. For example:

- Our November 2011 report on the Emergency Relief program (which provides funds to states to repair roads damaged by natural disasters and catastrophic failures) found that many of the project files reviewed did not contain documentation to support FHWA decisions that projects met program eligibility requirements.[38] Specifically, of the 83 projects reviewed, 81 projects (representing $193 million in federal funds) had missing or incomplete documentation. As a result, we were unable to determine the basis of FHWA's eligibility decisions for many of the projects reviewed. We also found that FHWA divisions relied heavily upon the information provided by states to make FHWA eligibility decisions without verifying that information. For example, one FHWA division office reported that it reviewed preliminary cost estimates for about one-third of the projects included in our review before determining that projects were eligible.[39] As a result, we could not determine the basis of FHWA's eligibility decisions for those project cost estimates it did not review and as such, FHWA ran the risk of providing funds to ineligible projects.

- In the Disadvantaged Business Enterprise (DBE) program, which aims to increase the participation of small businesses owned and controlled by socially and economically disadvantaged individuals, state DOTs are among those entities responsible for certifying firms to participate.[40] In an interview with one FHWA division office, the FHWA official said that he knows that the state DOT official is very experienced with the DBE certification process and, because of that, relies on the state to make certification decisions consistent with federal regulations. As a result, the official stated that FHWA is generally not involved in verifying the eligibility of DBE firms certified by the state. Although FHWA is not required to review every certification, in this instance FHWA's partnering relationship with the state influenced the level of oversight conducted in this area and exposed FHWA to the risk that ineligible firms might be certified as DBEs.
- Officials from the FHWA division offices we spoke to said they tended *not* to do unannounced inspections. Instead, FHWA alerts the relevant construction sites and offices of an inspection ahead of time. Officials from one FHWA division office explained that they rely primarily on announced visits because they do not want to create a "gotcha" environment, which might hurt their relationship with the state. Division officials from another office explained that announcing inspections gives state DOT staff time to do things like assemble the appropriate records or personnel for FHWA's inspection or allows FHWA to observe specific activities, such as materials testing, on the day that particular activity is occurring. While there are some advantages to announced inspections, the Institute of Internal Auditors includes unannounced visits as a common practice used by firms to mitigate risks associated with partnering.[41] By not conducting unannounced inspections, FHWA is essentially trusting the state and its contractors to put compliance with federal requirements over meeting competing demands like cost and schedule. In doing so, it may be missing the opportunity to more accurately verify compliance with federal requirements, observe normal operations, and create an environment conducive to compliance.

The partnering relationship between FHWA and state DOTs at times may have also resulted in FHWA being reluctant to require corrective action to bring a state back into compliance with program requirements. Specifically,

FHWA staff acknowledged that, in their daily decision making, they have to think about how to preserve their relationship with their state counterpart and that they view taking corrective action as potentially damaging to that relationship. For example:

- We and the DOT Office of Inspector General have reported multiple times on the problem of funds committed to inactive federal-aid highway projects.[42] FHWA has made it a priority to decrease nationally the number of outstanding inactive projects to ensure that federal funds are being used in a timely and effective way. For example, FHWA had reduced the percent of funds obligated to inactive projects to about 3.4 percent of all obligations by March of 2012—this percentage had stood at around 8 percent as recently as September 2010.[43] In 2008, it established a Financial Integrity Review and Evaluation program requiring division offices to conduct a quarterly review of inactive projects and determine the validity of the amount obligated for each project. FHWA division offices have the authority to de-obligate funds from inactive projects. However, FHWA division officials with oversight responsibility for three states we visited expressed reluctance to use this authority because of concerns that it would negatively affect their working relationship with the state. Instead, these division offices negotiated with state officials to get them to explicitly agree to allow FHWA to de-obligate funds. FHWA officials acknowledged that this is a long, time-intensive process. For example, over the course of 6 months, one FHWA division office sent reminder letters with specific deadlines for the state to provide a rationale for allowing inactive funds to remain obligated. Yet at the end of this process there were still outstanding inactive projects that had not been resolved. In another state, FHWA finance personnel described having ongoing conversations with their state counterparts, asking them the status of inactive projects and negotiating to de-obligate those funds. The FHWA division office described the process as "walking the tightrope" with the states when making decisions to de-obligate. The amount of time officials we spoke with devoted to addressing inactive funds raises questions about whether, on the whole, division offices could have moved more quickly to make these funds available to other needed projects had officials not had to consider the impact of withdrawing funds on their partnership with the state.

- One state identified serious compliance issues with one of its major cities dating as far back as 2003, including federal construction specifications not being followed, insufficient field equipment, and lack of appropriate construction supervision. In 2009, FHWA withheld funding from the city for about 2 weeks while the state DOT drafted a corrective action plan. FHWA approved the plan and resumed funding. However, nearly 2 years later, as of August 2011, there were still points in the plan that had not been addressed. As a result, federal funds continued to flow to projects that may not have fully met federal requirements.
- FHWA can delegate to the state the responsibility of approving consultant contracts to ensure compliance with federal regulations. One state DOT lacked FHWA-approved written procedures for how it selects consultants, which are necessary to comply with federal regulations. The FHWA division office had given the state an extended opportunity—about 5 years—to address the compliance issue, allowing it to use interim procedures as long as the state DOT was developing final procedures and planning to have them approved by FHWA. After 5 years, due to the failure of the state to develop final procedures, the FHWA division office suspended all state DOT contract approvals and temporarily re-assumed the responsibility of approving consultant contracts to ensure compliance with federal procurement regulations. Once the state DOT developed written procedures and they were approved, FHWA restored consultant contract approval authority to the state DOT.

As discussed earlier, FHWA division administrators reported that they rarely use their most punitive corrective action tools such as cutting off funding for a program or organization. While FHWA, in cooperation with the Federal Transit Administration, is responsible for certifying that MPOs meet federal requirements, as we reported in September 2009, FTA and FHWA officials were unaware of any instance in which an MPO was not certified due to noncompliance in the previous 10 years.[44] In our survey of division administrators, for fiscal years 2009 through 2012, one FHWA division administrator reported that he withheld certification of an MPO due to issues in its congestion mitigation plan. However, FHWA still allowed project approvals to move forward. The FHWA division office put together a corrective action plan with the MPO, identifying action steps, deadlines, and people responsible. According to the FHWA division administrator, as long as the state is making progress toward resolving the issue, FHWA will not prevent the MPO from obtaining project approvals and moving construction forward. He noted that MPOs have a 4-year window to become recertified, and if an MPO reached the end of its window but still had not taken sufficient action for FHWA to certify it, FHWA would likely provide an extension and grant the MPO "conditional" certification, rather than decertify it.

Similarly, none of the 52 FHWA division administrators stated they had not approved their state's bridge program any time in the past 3 fiscal years. In 2010, the DOT Office of Inspector General found some cases where FHWA bridge engineers reported that a state's bridge program substantially complied with federal regulations despite deficiencies that could have posed serious risks to public safety.[45] For example, one FHWA bridge engineer judged a state to be substantially compliant despite reporting that the state failed to close 96 bridges, as required. A bridge engineer in another state reported that 47 bridges were not closed as required, but concluded that the state was substantially compliant. In two other cases, FHWA bridge engineers reported states as substantially compliant even though 200 bridges in one case and over 500 bridges in the other case were not posted with maximum weight limit signs, as required.

FHWA Sometimes Showed a Lack of Independence in Decisions

FHWA's partnership relationship with the state may have made it vulnerable to a lack of independence in decisions to advance the program in some cases, resulting in it putting the states' interests above federal interests. For example:

- The Recovery Act required that projects from economically distressed areas be given priority for project selection. The Public Works and Economic Development Act of 1965, as amended, defined these areas, and FHWA supplemented it with guidance to address determination questions. However, as we reported previously, three states developed their own guidelines to determine eligibility.[46] FHWA eventually determined that the states' efforts to determine eligibility did not meet the federal criteria, but it did so after obligating nearly all the funds in those states, meaning FHWA's determination was too late to affect project selection.[47]

- Our November 2011 report on the Emergency Relief program found instances in which two division offices funded new emergency events with allocations that were no longer needed from previous emergencies.[48] FHWA's procedures specify that FHWA division offices are to identify, and then withdraw, unused program funding allocations so that FHWA can direct these funds to other emergencies. This provision exists because Emergency Relief program funding is often insufficient to fund emergency needs and a "backlog" list of projects exists. These two FHWA division offices gave priority to the state's interest in maintaining control of funds over the federal interest of ensuring that scarce unused Emergency Relief funds were made available to other states with potentially higher-priority emergencies. In response to our recommendation that it do so, FHWA instructed its division offices in January 2012 to no longer permit states to transfer unobligated allocations from one emergency event to a new event.
- We have previously expressed concerns about the Emergency Relief program prioritizing state's interests in other instances as well.[49] In 1996, we questioned the use of Emergency Relief funding in relocating and bettering the Cypress Viaduct, an elevated highway that collapsed in a catastrophic earthquake in Oakland, California. FHWA's regulations limit the use of emergency relief funds for improvements to or changes in the character of a destroyed facility. The regulations allow for funding "betterments"—such as relocation, replacement, upgrades, or added features that did not exist prior to the disaster—only when they are clearly economically justified to prevent recurring damage. In 1991, FHWA and the California Department of Transportation decided, as a result of environmental and community concerns, to replace the collapsed 1.5 mile connection with roughly 5 miles of new freeway segments. We questioned funding these improvements and additional costs through the Emergency Relief program, rather than through the annual formula funding states receive from the federal-aid highway program. This decision provided California with over $1 billion in additional funding that it then did not have to utilize from regular federal-aid funds or state sources. In 2007 and 2011, we reported additional cases of FHWA using the Emergency Relief program to fund projects that had grown in scope and cost as a result of environmental and community concerns. In 2007, we recommended that FHWA revise its regulations to tighten eligibility criteria and place limits on the use of Emergency Relief

program funds to fully finance projects with scope and costs that have grown as a result of environmental and community concerns. FHWA has not acted on this recommendation.

In addition, FHWA has on occasion taken extraordinary measures, expending considerable resources to advance the program, to the point of becoming actively and closely involved in developing and implementing solutions to state problems. When an overseer becomes part of the solution, the arms-length, independent perspective may be lost, as agencies that are responsible for implementing program improvements face an inherent conflict when they later approve those actions or review their effectiveness. For example:

- FHWA spent a substantial amount of time and effort with state DOT personnel and others trying to determine if funds used on a private bridge were eligible for use to help the state meet its matching requirement for federal funding. According to division office officials, the state was struggling to meet the 20 percent funding match required of states in order to receive the 80 percent federal-aid highway funding, due to the economic recession and poor fiscal situation of the state. Division office officials further explained that had it not met the match, the state could have lost about $200 million in federal funds in fiscal year 2012. States may receive "toll credits"—funds that can be credited by FHWA toward the state's federal match—if it can demonstrate that toll revenues were spent on facility improvements and meet other requirements. In an effort to meet its match, officials explained that the state identified a private toll bridge that it repaired and improved using toll revenue, but had never claimed the revenues as federal toll credits. The FHWA division office committed staff, including their financial manager, to work with the state DOT, the private bridge company, and an outside auditing firm to determine the eligibility of the toll credits. Together they identified more than $50 million in eligible toll expenditures, finding individual line item expenditures in areas such as preventative maintenance and capital improvements related to tolling equipment and real estate acquisition. The eligible toll credits helped the state meet its fiscal year 2012 federal match requirement. While FHWA officials characterized the Division's activities as appropriate technical assistance that was needed because the state did not have the skills to identify these credits, it placed the agency in a position of approving actions it was actively and closely involved in developing.

- In one state, an FHWA division office detailed a staff person to work full-time on-site at the state DOT to help bring the state into compliance with the requirements of the Highway Beautification Act of 1965.[50] Division office officials had identified ongoing compliance issues with the state's outdoor advertising program, including multiple examples of signs that were not in compliance with the state's agreement with DOT.[51] As a result, FHWA could have withheld 10 percent of the state's federal-aid funds. However, the division did not withhold these funds. The division adopted an approach that entailed considerable time and effort on the part of FHWA by detailing a staff person to (1) research sign regulations on federally controlled routes to determine compliance with the federal-state agreement and the Highway Beautification Act, (2) review the state's outdoor advertising inventory to determine the status of signs, and (3) provide interpretations, clarifications, and authoritative determinations concerning FHWA policy, among other activities.

Despite Partnership Risks, FHWA Has Good Oversight Practices in Several Areas

Despite the risks partnership poses, in several areas of the federal-aid highway program, FHWA has good oversight practices. The Institute of Internal Auditors has identified segregation of duties in the type of partnership FHWA has with state DOTs as one of the most common practices used in managing partnership-related risks.[52] During the administration of the Recovery Act, FHWA developed the National Review Teams (NRT) composed of FHWA staff—separated from the rest of the FHWA—to act as a neutral third party to conduct oversight. The NRTs were able to maintain their neutrality and objectivity in part because they did not have to concern themselves with preserving a partnering relationship while conducting oversight and making recommendations for action. In addition, the findings and recommendations of the NRTs were reported both to the FHWA division office, which was responsible for developing action items in response, and the responsible FHWA Director of Field Services, who was responsible for ensuring the action items were completed within the established time frames.

This practice of providing an independent review had several benefits. According to FHWA officials, it provided

- a consistent, comparative perspective on the oversight regularly conducted by division offices, and it gathered information at the national level on both best practices and recurring trouble spots across FHWA division offices;
- additional "boots on the ground" for project-level oversight and increased awareness of federal oversight activity among states, MPOs, and other transportation organizations receiving Recovery Act funds; and
- an independent outside voice to examine the Recovery Act projects and point out problems, keeping the partnering relationship between the division offices and the state DOTs intact.

The response to the NRT reviews from both division office and state officials with whom we spoke was positive. For example, division office officials said that the NRT reviews often echoed their own observations of weaknesses in the state DOT's program, but they said the state DOT seemed more inclined to act because the NRT was a fresh voice presenting the observations. Division office officials also told us that having the NRT point out deficiencies was helpful to them in maintaining their partnering relationship with the state. State DOT officials we spoke with in our discussion groups generally agreed that the NRT reviews, while they created an additional burden, provided an independent third-party perspective during the implementation of the Recovery Act.

In administering the federal-aid highway program, FHWA makes use of two practices that facilitate good oversight:

- A risk management approach to oversight. Conducting risk assessments, which are part of FHWA's approach, to identify both internal and external risks to an agency is another best practice for agencies.[53] In particular, by targeting areas of risk at both the state and national level, FHWA can focus on specific program areas of concern and better utilize limited resources. Division offices conduct annual assessments of their states to identify the greatest risks and vulnerabilities, and FHWA headquarters uses this information to identify common risk areas across the nation. FHWA officials in several division offices we spoke with stated that they use the risk assessment to inform their oversight activities throughout the year and take specific steps, when applicable, to address the risks. In 2009, we reported that FHWA had improved its use of risk assessments by proactively identifying risks and their potential impact, as well as developing specific response strategies to inform its planned oversight activities. Our report concluded that FHWA's guidance and training reflected best practices in risk management in three of four key areas.[54]

- Random sampling to review documentation of various financial transactions—a practice that is in keeping with its risk-based approach. This approach ensures that FHWA can assess compliance with financial requirements in a systematic way when it is impossible for FHWA to survey all occurring financial transactions. Furthermore, selecting transactions randomly protects the sample from selection bias, to which FHWA division offices could potentially be vulnerable because of their partnering relationship with the state. For example, in its Financial Integrity Review and Evaluation system, FHWA headquarters selects a random sample of transactions for each division office to check for compliance with the Improper Payments Information Act of 2002.[55] FHWA headquarters also randomly selects billing transactions each quarter to determine if there is sufficient documentation to support the billing item and amount.

PARTNERING COULD HELP FHWA IMPLEMENT A PERFORMANCE-BASED SYSTEM, BUT LONG-STANDING CHALLENGES WOULD REMAIN

Legislation has been approved in the Senate that would move the federal-aid highway program toward a more performance-based approach. FHWA's partnership—its close working relationship with the states—could be useful in making the transition to such a system; however, FHWA would need to effectively address the risks posed by such a close partnership—lax oversight and lack of independence. In addition, it would have to address other existing weaknesses that we have identified in previous reports, including improving the transportation planning process and data collection and evaluation. Finally, long-standing challenges stemming from the growth in the number of responsibilities and complexity within the federal-aid highway program, as well as the lack of well-defined federal goals and roles, would remain. Reexamining and refocusing surface transportation programs, which we have previously recommended, presents an opportunity to narrow the scope of FHWA's responsibilities so that it is better equipped to transition to a performance-based system. This review identified areas where FHWA expends considerable time and resources but exercises little effective control—areas where devolving responsibilities to the states may be appropriate.

Legislation Approved by the Senate for a Performance-Based Highway Program

A performance-based system is critical to the reexamination and restructuring of surface transportation programs that we and others have recommended.[56] Currently, most highway grant funds are distributed through formulas that have only an indirect relationship to infrastructure needs and many have no relationship to outcomes or the performance of the grantees. Because funds are distributed without regard to performance, it is difficult to know whether federal spending is improving the performance of the nation's highway infrastructure.[57]

Under Moving Ahead for Progress in the 21st Century (MAP-21)—FHWA would develop performance targets for minimum condition levels in two areas: (1) pavement on the Interstate and non-Interstate highways on the National Highway System and (2) bridges on the National Highway System.[58] If a state did not to meet the minimum condition levels for 2 consecutive years, it would be required to commit a specific percentage of its federal-aid highway funding to the deficient area. For other areas, MAP-21 directs states to develop performance targets related to national priorities identified in the legislation,[59] document these targets in their statewide transportation improvement programs, and link investment priorities to these targets. FHWA would have to (1) work with the states to develop performance goals that represent real improvements relative to the state's current conditions and will improve the performance of the nation's transportation system, and (2) monitor and measure states' progress and take corrective action should states not meet performance targets. Legislation approved by the House Transportation and Infrastructure Committee and the President's fiscal year 2013 budget proposal also refer to states developing performance measures and goals to improve safety, congestion, and other areas.

FHWA's Partnership Approach Could Facilitate a Transition to a Performance-Based Program

FHWA's partnership with states could offer several benefits in moving toward a performance-based program. In particular, through the partnership's collaborative approach, FHWA could provide technical assistance to help states develop performance goals and targets and establish data collection methodologies to evaluate and track their progress. States participating in our discussion groups found the technical assistance, knowledge transfer, and policy advice that FHWA provides a highly valuable benefit of the partnership relationship. Likewise, FHWA division office personnel recognized the benefit partnership offers in facilitating technical assistance.

Developing effective performance goals and targets and the data collection methods to track targets poses challenges that technical expertise can address. As we have reported, the more specific, measurable, achievable, and outcome-based the goals are, the better the foundation for allocating resources and optimizing results.[60] Also, goals must be linked to project selection and funding decisions, and without specific and measurable outcomes for federal involvement, policymakers will have difficulty determining whether certain parts of the federal-aid highway program are achieving the desired results. In addition, developing data collection methods that consistently and reliably capture the metrics needed requires technology, planning, and training staff to ensure high-quality data.

FHWA has recently developed performance metrics and revamped its data collection approach for the National Bridge Inventory System. Specifically, it adopted a new risk-based, data-driven approach that incorporates the review of 23 individual performance metrics, and, where appropriate, makes use of random sampling of the state's bridges to evaluate the metrics.[61] This recent experience, coupled with its technical expertise in other areas and division office officials' relationship with, and knowledge of, their state, would help to facilitate knowledge transfer and have the potential to create an effective performance-based program.

FHWA Would Need to Mitigate Risks of Partnership

Moving to a more performance-based approach means monitoring and measuring states' progress, holding states accountable for meeting performance targets, and taking corrective action objectively and consistently across states when needed. However, this can only be achieved if the risks posed by partnership discussed earlier—lax oversight, reluctance to take corrective action, and lack of independence in decision making—are overcome.

In addressing the risks posed by its partnership, FHWA can draw on some of its existing organizational structures. For example, during the implementation of the Recovery Act, FHWA used its NRTs to augment the oversight provided by division offices by conducting additional programmatic reviews and project inspections. Officials stated that the NRTs were able to maintain their neutrality and objectivity while conducting oversight and making recommendations, and that NRT personnel provided a consistent, comparative perspective to the oversight regularly conducted by division offices. Officials also explained that the NRTs' observations often reinforced those of division staff while also allowing the partnering relationship between the division offices and state DOTs to remain strong.

FHWA Would Need to Address Existing Oversight Weaknesses in Planning and Data Collection

Although FHWA would have to work with the states to develop performance goals and monitor and measure states' progress, we have reported weaknesses in federal oversight of both the statewide and metropolitan area planning processes that prevent effective measurement and tracking of performance outcomes. For example, we found that FHWA's oversight of statewide planning focuses on process, rather than specific transportation outcomes.[62] As such, FHWA cannot assess whether states' investment decisions are improving the condition and performance of the nation's transportation system. Similarly, pursuant to federal law, federal oversight of metropolitan planning is process-oriented rather than outcome-oriented, making it difficult to determine whether this oversight was improving transportation planning. Specifically, FHWA's oversight is geared toward determining whether MPOs are in compliance with federal laws and regulations, and this procedural focus, coupled with the fact that FHWA rarely withholds certification of MPOs, makes it difficult to use the certification process as a performance indicator for MPOs. In addition, we found that while FHWA identifies corrective actions to bring MPOs into compliance, it does not routinely assess the progress MPOs are making toward completing those corrective actions.

We recommended to Congress that FHWA more closely review states' transportation improvement programs to assess whether states' investments are achieving intended outcomes, rather than limiting its evaluation to whether the state complied with federal processes for developing the plan. We also recommended that Congress make the metropolitan planning processes more performance-based in order for FHWA to better assess the MPOs' progress in achieving results and better understand whether federal funds are being used to achieve national goals.[63] When we completed our review, Congress has not yet approved a multi-year surface transportation reauthorization measure, which could potentially address these recommendations.

In addition, FHWA would need to improve its ability to collect national-level data on highway performance to hold states accountable. Although the Government Performance and Results Act of 1993[64] requires agencies to measure performance toward the achievement of program goals and objectives,[65] we have stated previously that the federal government is not equipped to implement a performance-based approach to transportation funding because it lacks comprehensive data.[66] For example, during the administration of the Recovery Act we recommended that DOT assess the long-term benefits of Recovery Act investments in transportation infrastructure. In its response, DOT said it expected to be able to report on Recovery Act outputs, such as the miles of road paved, but not on outcomes, such as reductions in travel times. We have found other areas in which the lack of comprehensive, national- level data would hinder any move toward a performance-based system. For example, in administering the on-the-job training program, FHWA does not collect consistent national-level data on the number and demographics of program participants, the trades involved, and status of trainees. As a result, FHWA is not able to assess program results and hold states accountable.[67] In another example, our review of statewide transportation planning found that while FHWA division offices were collecting data on progress made by states to advance projects on their statewide transportation improvement programs to their construction phase, the data were unreliable to the point of being unusable because they were collected inconsistently across states and could not be used to compare states' progress.[68] We have made numerous recommendations to DOT related to the need for national-level data—a number of which DOT has yet to implement.

Long-standing Challenges and the Need to Reexamine and Reform the Federal-Aid Highway Program Remain

FHWA's success in transitioning to a performance-based program is dependent not only on addressing risks posed by its partnering relationship, but also by factors it cannot control. A performance-based program represents new responsibilities at a time when the growth in the number and complexity of its responsibilities and the lack of well-defined federal goals and roles leave FHWA, to a large extent, with a broad mandate in an increasingly constrained budget environment. As we have reported previously, a performance-based system is one part of the broader need to reexamine and restructure the program. A clearer definition of the federal role and, in turn, FHWA's responsibilities is under Congress's purview, and therefore beyond FHWA's or its partners' ability to address.

In 2008, we recommended that Congress consider a fundamental reexamination and reform of surface transportation programs that would potentially result in a more clearly defined federal role in relation to other levels of government and thus a more targeted federal role focused around evident national interests. For issues in which there is a strong national interest, ongoing federal financial support and direct federal involvement could help meet federal goals. Where national interests are less evident, other stakeholders could assume more responsibility, and some programs and activities may better be devolved to other levels of government. In some cases, it may be appropriate to "turn back" activities and programs to state and local governments if they are best suited to perform them.[69]

Devolving parts of the federal-aid highway program may be appropriate in cases where not only national interests are less evident, but also where FHWA expends considerable time and resources yet exercises little effective control. During this review, we identified two such areas: (1) delegated projects and (2) locally administered projects. By law, the states assume oversight responsibility for delegated projects—that is, all projects that are not on the National Highway System, and are thus off the federal-aid system and not the focus of the national program.[70] In fiscal year 2011, about 48 percent of federal-aid funds were obligated for projects for which oversight could be assumed by the states. Even though the states are responsible for oversight of the design and construction of these projects, FHWA is still expected to evaluate state DOT capacity through a number of processes and reviews. Division office officials told us that these activities require a considerable amount of time and effort on the part of their staff.

Locally administered projects are projects in which a state DOT has given approval to a local public agency (e.g., a city or county) the responsibility to administer a project or phase of a project such as design, property acquisition, or construction. These projects can either receive full oversight from FHWA or that responsibility can be assumed by the state. During our review, FHWA did not have national-level information on the number of projects or amount of federal funds spent on locally administered projects. However, it began requiring division offices to collect this information for newly authorized projects on March 12, 2012, so such data will be available in the future. Nevertheless, locally administered projects are ranked by FHWA's risk assessments as among the highest-risk areas in FHWA's oversight portfolio at both the state and national levels. For example, at least 33 division offices included risks related to locally administered projects as their most pressing risk areas in 2010. These risks included a lack of understanding of federal-aid construction contract requirements and use of innovative or nontraditional construction techniques by inexperienced local agencies. Likewise, FHWA headquarters identified locally administered projects as high risk. According to FHWA division and state officials, local agencies struggle to meet federal regulations that accompany federal-aid funding because of high staff turnover at the local level and the infrequency with which local agencies receive federal funding.

These challenges were reiterated throughout the discussion groups we conducted, as well as at the site visits to FHWA division offices that we conducted across the country. FHWA officials from two states described a wide range of risks posed by locally administered projects, including use of outdated design standards, lack of quality control and assurance, lack of standard documentation and recordkeeping, and insufficient knowledge of the right-of-way acquisition requirements. One FHWA division office provided examples of locally administered projects in their state that did not conduct construction inspections or materials testing or bought supplies from foreign countries, actions that are out of compliance with federal regulations. According to officials in three of the FWHA division offices we visited, locally administered projects require considerable time, attention, and resources. For example, according to officials in three division offices, FHWA staff expend a good deal of time and effort providing technical assistance and capacity-building to enhance the ability of local agencies to successfully administer federal-aid projects.

Further, our analysis of the 2010 state Single Audits[71] showed that insufficient monitoring of subrecipients,[72] such as on locally administered projects, was one of the most common findings and that 18 of 47 reporting states had findings related to monitoring of subrecipients. Our analysis also showed that state DOTs did not properly communicate federal requirements in their awards to or contracts with subrecipients and that their monitoring of subrecipients during the award was inadequate. Specifically, state DOTs' monitoring of subrecipients for compliance with federal and state requirements lacked procedures for or had poor compliance with existing procedures for regular site visits, risk assessments, and performance reporting.

As we have reported, devolving parts of the federal-aid highway program would have many implications and would require careful decisions to be made at the federal, state, and local levels. Since the federal-aid highway program has a dedicated source of funding (in that it is funded from fuel taxes and other fees deposited into the Highway Trust Fund), devolving parts of the highway program could entail reducing revenues into the Highway Trust Fund. The decision to reduce revenues at this time would be difficult because the Congressional Budget Office estimates, as of March 2012, that to maintain current spending levels plus inflation between 2013 and 2022, the Highway Trust Fund will require over $125 billion more than it is expected to take in over that period.[73] At the federal level, it would need to be determined what functions would remain and how federal agencies would be structured and staffed to deliver those programs. At the state and local levels, it would need to be determined whether to replace federal revenues with state taxes and what types of programs to finance. Deciding whether to replace federal revenues with state taxes would be difficult because states also face fiscal challenges and replacing revenues would have different effects on different states.[74]

CONCLUSIONS

FHWA's partnership approach with the states allows it to proactively identify issues before they become problems, achieve cost savings, and gain states' commitment to improve their processes. In some areas, FHWA division offices have good oversight practices that complement its partnership, including using a risk-based approach to its oversight and using an independent, third- party review (the NRTs) to augment its oversight activities during the implementation of the Recovery Act.

However, FHWA's partnership also poses risks that it has not to date directly addressed, that can potentially result in improper or ineffective use of federal funds and the loss of independence necessary for effective oversight. Should Congress direct FHWA to move to a performance-based system, holding states accountable for achieving performance measures—and taking action when they do not—would be essential. Because of the nature of their partnership with the states, FHWA's division offices may not be in the best position to mitigate partnership risks. Given that partnership produces benefits, the solution does not lie with eliminating FHWA's partnership approach. Rather, a strategy built around leveraging the strengths of the partnership approach while managing its risks could provide a better way for FHWA to verify the states' use of federal funds.

While such a strategy could take many forms, greater separation of the responsibilities to advance, oversee, and make corrective action decisions in the program would be consistent with good internal control practices and may help FHWA transition to a performance-oriented program. Specifically, a nationally focused, independent oversight entity modeled on the NRTs could be an effective vehicle to mitigate risks associated with partnering between FHWA division offices and state DOTs by conducting periodic evaluations of selected activities and making recommendations for improvement. This could be particularly helpful in instances in which the division offices have been reluctant to take corrective action because of concerns about damaging the partnering relationship. In addition, if Congress directed FHWA to move to a performance-based system, an entity modeled on the NRTs could assess states' progress toward performance measures and hold states accountable for meeting them. Responsibilities such as technical assistance and knowledge transfer—areas where FHWA's partnering relationship can help states develop performance goals and targets—could remain with the division offices.

Any successful transition to a performance-based system in the highway program requires accurate, reliable national-level data. The partnership that division offices have with state DOTs could help to ensure that states develop data collection methods that would help determine whether the highway system overall was improving. Furthermore, FHWA has the expertise to develop and implement a rigorous national-level data collection effort as it recently did with the National Bridge Inventory System. We are not making a new recommendation to DOT on this matter because many of our recommendations on collecting national-level data remain open.

In 2008, we recommended that Congress consider reexamining and refocusing surface transportation programs, establishing well-defined goals with direct links to identified federal interests and roles, and consider devolving to the states and other levels of government responsibility for programs where national interests are less evident. The information we gathered during the course of this review and the pending transition to a more performance based federal-aid highway program reinforces the need to act. First, FHWA's responsibilities have expanded over the years while its resources have not, and the addition of a performance-based system to its already broad mandate would further expand FHWA's responsibilities. Reexamining and refocusing surface transportation programs presents an opportunity to narrow the scope of FHWA's responsibilities so that it is better equipped to transition to a performance-based system. Second, this review has identified specific areas where devolving or turning back to the states the responsibilities for managing and funding some parts of the highway program may be appropriate. Turnback would have many implications and would require careful decisions. Yet nearly half of federal-aid highway funds are spent on roads off the National Highway System—projects for which oversight has been assumed by the states—raising questions about whether evident federal interests are at stake. In addition, the considerable federal resources FHWA expends overseeing locally administered projects—including capacity-building activities for city and county governments—raises questions about whether such time and effort is better spent in support of more nationally focused programs and objectives.

MATTER FOR CONGRESSIONAL CONSIDERATION

As we have previously recommended, Congress should consider reexamining and refocusing surface transportation programs, including establishing well-defined goals with direct links to identified federal interests and roles. Based on this review, there may be areas where national interests are less evident and where Congress may wish to consider narrowing FHWA's responsibilities.

RECOMMENDATIONS FOR EXECUTIVE ACTION

We recommend that the Secretary of Transportation direct the FHWA Administrator to develop a strategy based on the NRT model to mitigate the risks associated with its partnering approach with state DOTs, while maintaining the strengths that the partnership approach brings to the program. This strategy should address existing risks and, if Congress directs FHWA to move to a performance-based system, partnering risks that could affect the successful implementation of such a system.

AGENCY COMMENTS

We obtained oral comments from DOT officials, including the Director of FHWA's Office of Program Administration. These officials stated that DOT generally agreed with the findings and recommendations in the report. Specifically, they recognized that the agency's partnership approach with the states poses oversight risks. They stated that they are implementing efforts based on the NRT model to provide independent reviews and accountability services to improve the efficiency and effectiveness of FHWA programs. We will monitor these efforts to assess if the department is responsive to our recommendation that DOT mitigate the risks of its partnership with the states.[75] DOT officials also provided technical comments which we incorporated as appropriate.

Sincerely yours,

Phillip R. Herr
Managing Director Physical Infrastructure Issues

APPENDIX I. FHWA ORGANIZATION

The Federal Highway Administration (FHWA) has 52 division offices—one in each state, the District of Columbia, and Puerto Rico—to carry out the day-to-day activities of the federal-aid highway program. These offices are generally located in the same city as the state departments of transportation (state DOTs), which is usually the state capital. In addition, jointly with the Federal Transit Administration, the FHWA operates four metropolitan offices in Philadelphia, Pennsylvania; New York, New York; Chicago, Illinois; and Los Angeles, California, which are extensions of their respective division offices. FHWA division offices are organized geographically under three directors of field service who provide administrative supervision and leadership on strategic initiatives to their constituent division offices. FHWA headquarters provides leadership and policy direction for the agency, and FHWA's Resource Center, with five locations, provides technical expertise, guidance, and training to the states in areas such as air quality, civil rights, construction, environment, safety and bridges.

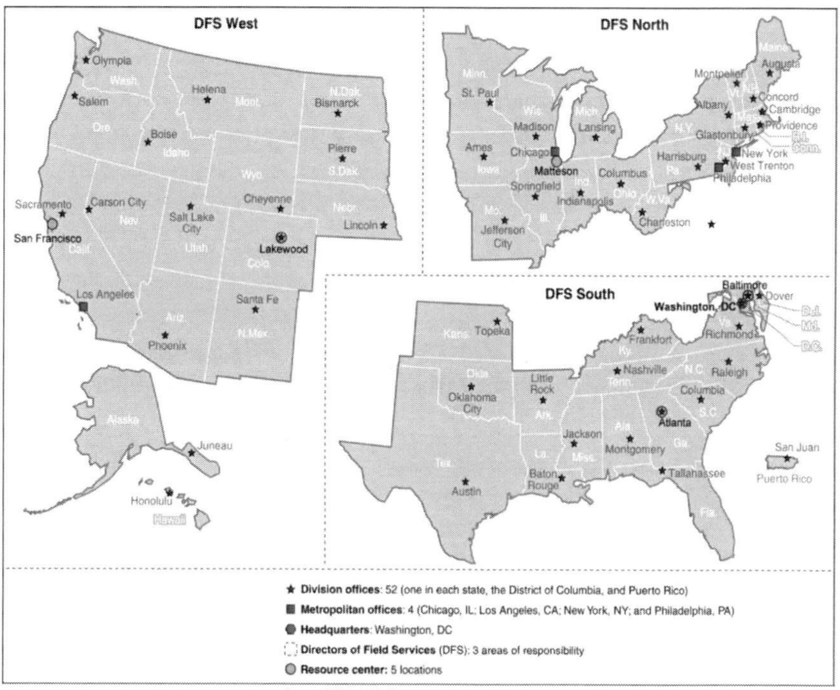

Sources: FHWA and Map Resources.

Figure 2. Locations of FHWA's Offices.

APPENDIX II. SCOPE AND METHODOLOGY

To address our objectives, we reviewed and analyzed relevant laws, regulations, and FHWA documentation. Specifically, we reviewed previous and current authorizations of the federal-aid highway program, as well as proposed reauthorization language. We also reviewed relevant sections of U.S. Code that pertain to FHWA and its relationship with the states. Additionally, we reviewed and summarized past GAO work regarding FHWA's programmatic oversight responsibilities and its relationship with the states.

To obtain information on current FHWA structure and oversight activities, we conducted site visits to nine FHWA division offices, including Colorado; Delaware/Maryland, which has a joint office; Maine; Michigan; North Carolina; Oklahoma; Virginia; and Washington, D.C. We selected these states to include a range of selection criteria, including the extent to which oversight responsibilities have been assumed by the state, the size of the state's transportation program and proportion of federal funding relative to state funds, the type of transportation system in the state (e.g., primarily rural highways and interstate or primarily infrastructure in densely populated urban areas); and geographic distribution. We interviewed FHWA and state DOT officials in multiple settings to learn how they characterize their relationship and what role partnership plays in oversight.

We also reviewed selected academic literature on formal partnering practices and tools. To perform this analysis, GAO conducted a variety of literature and Internet searches, reviewed previous GAO reports, and analyzed literature recommended by engagement stakeholders. We read, analyzed, and synthesized these documents to construct a common definition of partnering, namely that "partnering is an approach that, through collaborative processes and activities, enables parties to achieve individual and mutually beneficial goals and results." We also identified nine features of partnership from our review of the literature. Of those, we selected four features based on the applicability of these features to FHWA's partnering relationship with state DOTs based on our observations of FHWA, interviews with FHWA and state DOTs, and from our review of documentation of formal partnering arrangements between states and FHWA division offices. In doing so, we developed the following definition of features of partnership: "Partnering processes and behaviors span a continuum of collaborative activities including information sharing, participative and consultative processes, collaborative problem solving, and formal team-building such as charter signing and relationship assessment." Further, we reviewed literature to identify partnership risks, and after identifying a list of six risks, we identified two risks which were most evident in our audit work and that were most relevant to FHWA's partnering relationship with state DOTs: lax oversight and a lack of independence emerged as two primary themes of partnering risk.

To obtain an independent view of issues in FHWA's oversight, we examined the results of the 2010 Single Audits—statewide audits of financial statements and compliance with federal program requirements for certain programs among recipients of federal funds. Forty-seven states reported their results in the Federal Audit Clearinghouse as of October 28, 2011.[76] To determine relevant findings to our work, we identified audit records for FHWA funding categories, which provided us with funding amounts that were subject to audit findings as well as the types of audit findings.[77] We analyzed these findings to determine the types of findings occurring most frequently. For subrecipient monitoring, one of the most frequent audit finding types, we examined full-text Single Audit reports, comparing them against each other to identify common themes.

In addition to these efforts, we conducted a survey of all FHWA division administrators, who lead the FHWA division offices located in each state, as well as Washington, D.C., and Puerto Rico. With all 50 states and Washington, D.C., and Puerto Rico, our universe was 52 division offices.[78] We developed a web-based survey instrument of seven closed-ended questions and one open-ended question, regarding (1) FHWA's partnering relationship with state DOTs, and (2) FHWA's use of available corrective actions. We pre-tested the instrument with two division administrators in November 2011. The survey was released in December 2011. We received 52 completed surveys, for a 100 percent response rate. To obtain input from states on their relationship with FHWA division offices and their oversight of the federal-aid highway program, we conducted four discussion groups of state DOT representatives. We worked in conjunction with the American Association of State Highway and Transportation Officials to speak with personnel from a variety of geographic locations and various programs, including personnel from the areas of construction, locally administered projects, engineering, bridges, and leadership.

To determine the extent to which FHWA's incorporation of partnering practices into its oversight approach supports effective oversight, we used academic literature and GAO reports to identify criteria and effective practices for productive partnering and robust oversight. Using these criteria and effective practices, we assessed FHWA's current oversight practices by reviewing information from interviews with FHWA headquarters and division offices and state DOTs; site visit observations; and relevant findings from recent and ongoing GAO engagements examining various FHWA program areas.

To determine the extent to which FHWA's partnering approach serves as a foundation for moving toward a performance-based transportation program, we identified principles for a performance-based transportation system in previous GAO reports that can be applied to FHWA, including (1) national transportation goals, (2) performance measures, (3) appropriate performance targets, and (4) employing the best tools to emphasize return on investment. We also reviewed the proposed reauthorization bill, Moving Ahead for Progress in the 21st Century (MAP-21),[79] to incorporate Congress' expectations for moving toward a performance-based system.

We conducted this performance audit from April 2011 to April 2012 in accordance with generally accepted government auditing standards. Those standards require that we plan and perform the audit to obtain sufficient, appropriate evidence to provide a reasonable basis for our findings and conclusions based on our audit objectives. We believe that the evidence obtained provides a reasonable basis for our findings and conclusions based on our audit objectives.

End Notes

[1] Prior to 1967, the federal-aid highway program was administered by the Bureau of Public Roads, then a bureau in the Department of Commerce. These responsibilities were transferred to the Secretary of Transportation in 1967. Department of Transportation Act, Pub. L. No. 89-670, § 6(a), 80 Stat. 931, 937 (1966), eff. April 1, 1967, 32 Fed. Reg. 5453.

[2] GAO, *Federal-Aid Highways: FHWA Needs a Comprehensive Approach to Improving Project Oversight,* GAO-05-173 (Washington, D.C.: Jan. 31, 2005).

[3] GAO-05-173.

[4] S. 1813 §§ 1106, 1112, 112th Cong. (as adopted by the Senate March 14, 2012).

[5] FHWA's mission to advance the program is articulated in its national leadership strategic goal, which states that FHWA "leads in developing and advocating solutions to national transportation needs" as well as in its stewardship agreements.

[6] The National Highway System consists of approximately 160,000 miles of roadway important to the nation's economy, defense, and mobility. It includes Interstates, principal arterials, roads, and connectors important to the United States' strategic defense policy. These roadways provide access, continuity, and emergency capabilities for defense purposes, and connect to major intermodal facilities, such as airports or transit hubs.

[7] Federal and state governments oversee this regional planning process. At the federal level, the Federal Transit Administration and FHWA work together to perform federal certification reviews—certifying that each Transportation Management Area (TMA) has carried out its planning according to the applicable federal statutes. More specifically, the certification review requires that the federal government assess TMAs every 4 years to determine how well they are working with the transportation-related organizations, local governments,

public transportation operators, and citizens in their area, as well as with the state DOTs, to meet the many statutory and regulatory requirements applicable to the planning process. Additionally, the certification review assesses the quality of the required planning documents. The certification review includes a desk review of the MPO's plans and a site visit, among other things. Additionally, all MPOs, including both TMAs and non-TMAs, must also self-certify that their planning process meets the federal requirements.

[8] Required by 23 U.S.C. § 134(k)(5).

[9] Act of July 11, 1916, ch. 241, § 4, 64th Cong., 39 Stat. 355, 357.

[10] The act also referred to the federal-state relationship as contractual and authorized withdrawal of a portion of the San Antonio North Expressway as a federal-aid highway project upon repayment into the Highway Trust Fund of funds previously paid by the federal government. Texas' desire to withdraw the Expressway stemmed from an impasse over routing to minimize environmental impact by complying with the National Environmental Policy Act of 1969, Pub. L. No. 91-190, 83 Stat. 852 (1970) (codified, as amended at 42 U.S.C. ch. 55). Congress has enacted legislation authorizing states to withdraw highways from the program in other specific instances as well.

[11] FHWA policy documents and officials have altered this language slightly and widely refer to the program today as a "federally assisted state-administered program."

[12] GAO-05-173.

[13] Responsibility for public roads was at various times with the Department of Agriculture (Act of July 11, 1916, § 1, 39 Stat. 355), Federal Works Administration (Reorganization Plan No. 1 of 1939, § 302, 53 Stat. 1423, 1427), and the Department of Commerce (Reorganization Plan No. 7 of 1949, 63 Stat. 1070). The Department of Transportation Act established a Federal Highway Administration, headed by an administrator, into which was transferred the Department of Commerce Bureau of Public Roads reporting to its former administrator, retitled the Director of Public Roads. (Pub. L. No. 89-670, §§ 3(d), (4), 80 Stat., 932-933 (1066)). The position of Director of Public Roads was abolished by the Federal-Aid Highway Act of 1970, Pub. L. No. 91-605, § 114, 84 Stat. 1713, 1722-1723.

[14] Act of December 20, 1944, ch. 626, §§ 1, 3(b), 58 Stat. 838-839. Federal-aid-Secondary Program (1952-1991), Urban Extensions (1944-1976).

[15] Federal-Aid Highway Act of 1950, ch. 912, § 1(b), (c), 64 Stat. 785, 786 (1950), *See,* also, the Federal-Aid Highway Act of 1948, ch. 732, 62 Stat. 355 (1948).

[16] Act of December 20, 1944, § 8, 58 Stat., 842.

[17] Federal-Aid Highway Act of 1956, ch. 462., Title I, 70 Stat. 374.

[18] Highway Revenue Act of 1956, ch. 462, Title II, 70 Stat. 387.

[19] Highway Safety Act of 1970, 91-605, Title II, § 204(a), 84 Stat. 1739, 1741 (1970).

[20] Intermodal Surface Transportation Efficiency Act of 1991 (ISTEA), Pub. L. No. 102-240, § 1007(b)(1), 105 Stat. 1914.

[21] ISTEA, Pub. L. No. 102-240, § 1007(c), 105 Stat. 1914, listed these as making "provision of facilities for pedestrians and bicycles, acquisition of scenic easements and scenic or historic sites, scenic or historic highway programs, landscaping and other scenic beautification, historic preservation, rehabilitation and operation of historic transportation buildings, structures, or facilities (including historic railroad facilities and canals), preservation of abandoned railway corridors (including the conversion and use thereof for pedestrian or bicycle trails), control and removal of outdoor advertising, archaeological planning and research, and mitigation of water pollution due to highway runoff."

[22] Pub. L. No. 91-190, 83 Stat. 852 (1970) (codified as amended at 42 U.S.C. ch. 55).

[23] Pub. L. No. 87-866, § 9, 76 Stat. 1148 (1962).

[24] Congress authorized the final funding apportionments to complete the construction of the Interstate Highway legislation and created the National Highway System, consisting of the Interstate Highway System and other roadways connecting to major transportation hubs such as ports, airports, international border crossings, and public transportation and transit facilities. ISTEA, §§ 2, 1001, 1003(a)(2),1006, 105 Stat. 1914. DOT and the states

cooperatively identified these roads and selected 160,000 miles of interstate and other roads that by 1996 carried more than 40 percent of all highway traffic, 75 percent of heavy truck traffic, and 90 percent of tourist traffic. Once the roads were selected, Congress formalized this plan in the National Highway System Designation Act of 1995. Pub. L. No. 104–59, § 101(a), 109 Stat. 568 (1995).

[25] FHWA's risk management approach requires field offices to identify risks to the federal-aid highway program, assess the risks on the basis of the potential impact and the likelihood that they will occur, develop response strategies for key risks, and include these strategies in the offices' planned oversight activities.

[26] GAO, *Surface Transportation: Restructured Federal Approach Needed for More Focused, Performance-Based, and Sustainable Programs*, GAO-08-400 (Washington, D.C.: Mar. 6, 2008).

[27] GAO, *Surface Transportation: Principles Can Guide Efforts to Restructure and Fund Federal Programs,* GAO-08-744T (Washington, D.C.: July 10, 2008).

[28] GAO, *High-Risk Series: An Update*, GAO-11-278 (Washington, D.C.: February 2011).

[29] Pub. L. No. 111-5, Title XII, 123 Stat. 115 (2009).

[30] Administrative discretion is a public official's ability to decide whether a particular administrative act is to be implemented, and in what manner.

[31] 49 C.F.R. Part 26, Appendix A. When a state DOT sets Disadvantaged Business Enterprise contract goals on individual U.S. DOT-assisted contracts, bidders on those contracts must make good faith efforts to meet those goals. The bidder can meet this requirement in one of two ways: (1) meet the goal on the individual U.S. DOT-assisted contract, or (2) document that it made an adequate good faith effort—meaning that the bidder took the necessary and reasonable steps to achieve the goal even though it did not succeed in obtaining enough DBE participation to do so. State DOTs are responsible for evaluating whether bidders made good faith efforts to meet their goals, and according to U.S. DOT officials, this evaluation is subject to FHWA review as appropriate.

[32] 23 U.S.C. § 116(a).

[33] 23 U.S.C. § 116(d). Routine maintenance is defined as work that is planned and performed on a routine basis to maintain and preserve the condition of the highway system at an adequate level of service. Preventive maintenance is defined as a planned strategy of cost-effective treatments to an existing roadway system that preserves the system, retards future deterioration, and maintains or improves the functional condition of the system.

[34] GAO, *Transportation Infrastructure: Managing the Costs of Large-Dollar Highway Projects*, GAO/RCED-97-47 (Washington, D.C.: Feb. 27, 1997).

[35] GAO, *Federal-Aid Highways: FHWA Needs a Comprehensive Approach to Improving Project Oversight*, GAO-05-173 (Washington, D.C.: Jan. 31, 2005), 5.

[36] GAO, *Transportation: Central Artery/Tunnel Project,* GAO/RCED-95-213R (Washington, D.C.: June 2, 1995), *Transportation Infrastructure: Central Artery/Tunnel Project Faces Financial Uncertainties*, GAO/RCED-96-131 (Washington, D.C.: May 10, 1996), *Transportation Infrastructure: Progress On and Challenges to Central Artery/Tunnel Project's Costs and Financing,* GAO/RCED-97-170 (Washington, D.C.: July 17, 1997), and GAO/RCED-97-47 .

[37] FHWA, *Federal Task Force on the Boston Central Artery Tunnel Project: Review of Project Oversight and Costs* (Mar. 31, 2000).

[38] GAO, *Highway Emergency Relief: Strengthened Oversight of Eligibility Decisions Needed,* GAO-12-45 (Washington, D.C.: Nov. 8, 2011).

[39] FHWA told us that this risk-based approach was appropriate since states have assumed oversight responsibility for design and construction of many federal-aid highway projects, including emergency relief projects. However, we found that this approach did not appear to be consistent with FHWA's procedures as outlined in its Emergency Relief Manual.

[40] GAO, *Disadvantaged Business Enterprise Program: Assessing Use of Proxy Data Would Enhance Ability to Know if States Are Meeting Their Goals*, GAO-12-78 (Washington, D.C.: Oct. 13, 2011).

[41] Institute of Internal Auditors Research Foundation, *Managing Strategic Alliance Risk: Survey Evidence of Control Practices in Collaborative Inter-organizational Settings* (2006), 7.

[42] Inactive projects are state and local construction projects to which federal funds have been obligated, but which have not had any expenditures over a 1-year period or for which funds are available but are unlikely to be obligated by FHWA within 1 year, as certified by the state.

[43] FHWA tracks data on inactive projects and, although it does not publish these data, it provided them to us in commenting on this report.

[44] GAO, *Metropolitan Planning Organizations: Options Exist to Enhance Transportation Planning Capacity and Federal Oversight* GAO-09-868 (Washington, D.C.: Sept. 9, 2009).

[45] DOT, *Assessment of FHWA Oversight of the Highway Bridge Program and the National Bridge Inspection Program* (Washington, D.C.: Jan. 14, 2010), 9-10.

[46] GAO, *Recovery Act: States' and Localities' Uses of Funds and Actions Needed to Address Implementation Challenges and Bolster Accountability*, GAO-10-604 (Washington, D.C.: May 26, 2010).

[47] GAO-10-604 and GAO, *Recovery Act: Opportunities to Improve Management and Strengthen Accountability over States' and Localities' Uses of Funds*, GAO-10-999 (Washington, D.C.: Sept. 20, 2010).

[48] GAO-12-45.

[49] GAO, *Emergency Relief: Status of the Replacement of the Cypress Viaduct*, GAO/RCED-96-136 (Washington, D.C.: May 6, 1996).

[50] Highway Beautification Act of 1965, Pub. L. No. 89-285, Title I, 79 Stat. 1028 (1965) (codified as positive law in 23 U.S.C. § 131).

[51] The 1965 act requires each state DOT to enter into a formal agreement with DOT to control signs adjacent to the Interstate System and other routes included on what is now the National Highway System.

[52] Institute of Internal Auditors Research Foundation, *Managing Strategic Alliance Risk: Survey Evidence of Control Practices in Collaborative Inter-organizational Settings.*

[53] GAO, *Standards of Internal Control in the Government,* GAO/AIMD-00-21.3.1 (Washington: D.C. November 1999), 10.

[54] GAO, *Federal-Aid Highways: FHWA Has Improved Its Risk Management Approach, but Needs to Improve Its Oversight of Project Costs,* GAO-09-751 (Washington, D.C.: July 24, 2009).

[55] Pub. L. No. 107-300, 116 Stat. 2350 (2002).

[56] GAO-08-744T.

[57] GAO-08-400.

[58] S. 1813 §§ 1106, 1112.

[59] These priorities include (1) safety, (2) infrastructure condition, (3) system reliability, (4) freight movement and economic viability, and (5) environmental sustainability (S. 1813, § 1203).

[60] GAO-08-400.

[61] However, at the moment the National Bridge Inspection Standards' random sampling selection does not use a stratified approach that may be beneficial in this case. As a result, it is equally likely that a small, low-traffic, locally owned bridge would be selected for inspection as a large, high-volume, load-rated bridge on the Interstate.

[62] GAO, *Statewide Transportation Planning: Opportunities Exist to Transition to Performance-Based Planning and Federal Oversight*, GAO-11-77 (Washington, D.C.: Dec. 15, 2010).

[63] GAO-09-868.

[64] Pub. L. No. 103-62, 107 Stat. 285 (1993) (codified at 31 U.S.C. §§ 1115-1116).

[65] GAO, *Executive Guide: Effectively Implementing the Government Performance Results Act*, GAO/GGD-96-118 (Washington, D.C.: June 1996).

[66] GAO-08-744T.

[67] GAO-11-703.

[68] GAO-11-77.

[69] GAO-08-400.

[70] 23 U.S.C. § 106(c)(2).

[71] A Single Audit is a required audit of states, local governments, and nonprofit entities that expend at least $500,000 per year in federal awards. Single Audits include determinations on whether the audited entity met the compliance requirements listed in the Office of Management and Budget's Circular No. A-133 Compliance Supplement for each major program. There are 14 types of compliance requirements which include allowable costs/cost principles, activities allowed or unallowed, and subrecipient monitoring.

[72] A subrecipient is an entity that receives a grant award from the prime recipient of an award and is accountable to the prime recipient for the use of the federal funds provided by the subaward. For the federal-aid highway program, this is generally a local public agency that administers a federal-aid project, but may also be a nonprofit, educational institution, and in some cases, another federal agency.

[73] Congressional Budget Office-March Fiscal Year 2012 Baseline Projections for the Highway Trust Fund.

[74] GAO-08-400.

[75] Section 236 of the Legislative Reorganization Act of 1970 requires the head of the agency to submit a written statement on action taken on recommendations and submit the statement to the Committee on Homeland Security and Governmental Affairs of the Senate and the Committee on Oversight and Government Reform of the House of Representative within 60 days of the release of this report, and the Committees on Appropriations of both Houses of Congress in the first request for appropriations submitted more than 60 days after the date of the report. Pub. L. No. 91-510, § 236, 84 Stat. 1140, 1171 (codified as amended at 31 U.S.C. § 720(b)).

[76] Four of the remaining states do not submit annual Single Audit reports, and the fifth state did not submit any statewide Single Audit results for 2010 as of October 28, 2011.

[77] Single Audit finding types include activities allowed or unallowed; allowable costs/cost principles; cash management; Davis-Bacon Act; eligibility; equipment and real property management; matching; level of effort; earmarking; period of availability of federal funds; procurement and suspension and debarment; program income real property acquisition and relocation assistance; reporting; subrecipient monitoring; special tests and provisions; and other.

[78] One division administrator oversees both the Puerto Rico and Florida division offices. The survey instrument was sent out for each individual division office, and the administrator was given instructions on how to determine which instrument was assigned to which division office, and to respond fully to both, which he has done.

[79] S. 1813 §§ 1106, 1112, 112th Cong. (as adopted by the Senate March 14, 2012).

In: Highways
Editors: B. E. Allen and J. G. Gray

ISBN: 978-1-62257-733-0

Chapter 2

HIGHWAY PROJECTS: SOME FEDERAL AND STATE PRACTICES TO EXPEDITE COMPLETION SHOW PROMISE*

United States Government Accountability Office

WHY GAO DID THIS STUDY

Projects to construct, improve, and repair roads and bridges are fundamental to meeting the nation's mobility needs. However, completing highway projects—which generally involves four phases consisting of (1) planning, (2) preliminary design and environmental review, (3) final design and right-of-way acquisition, and (4) construction—can sometimes take a long time. In 2005, SAFETEA-LU established provisions to help expedite highway projects, including streamlining some portions of the environmental review process, allowing states to assume greater environmental review responsibilities under certain conditions, and establishing efforts that permitted delegation of some authority from the federal government to states. GAO was asked to (1) describe the process and factors that could affect highway project time frames, (2) examine state DOTs' views on the benefits and challenges of the provisions to expedite highway projects established in SAFETEA-LU, and

* This is an edited, reformatted and augmented version of the Highlights of GAO-12-593, a report to congressional requesters, dated June 2012.

(3) describe additional initiatives that state DOTs and FHWA have implemented to expedite the completion of highway projects. GAO surveyed officials from 52 state DOTs, including all states, the District of Columbia, and Puerto Rico; interviewed officials at FHWA, state DOTs, and federal resource agencies (agencies tasked with protecting natural, historic, or cultural resources); and analyzed legislation, regulations, and other reports and publications. U.S. DOT provided technical comments on a draft of this report, which GAO incorporated as appropriate.

WHAT GAO FOUND

The process to complete highway projects is complicated and lengthy due to multiple factors. Specifically, highway projects can involve many stakeholders, including agencies at all levels of government, nongovernmental organizations, and the public. These stakeholders perform a number of tasks—for major highway projects, as many as 200 steps from planning to construction—but their level of involvement varies. For example, resource agencies like the U.S. Army Corps of Engineers or the U.S. Fish and Wildlife Service generally only become involved in a highway project if it affects the environmental or cultural resources that agency is tasked with protecting. Additional factors can lengthen project time frames, including the availability of funding, changes in a state's transportation priorities, public opposition, or litigation.

State departments of transportation (DOT) that GAO surveyed generally agreed that the provisions meant to help expedite highway projects established in the Safe, Accountable, Flexible, Efficient Transportation Equity Act: A Legacy for Users (SAFETEA-LU) could decrease time frames but found some provisions more useful than others. They most frequently agreed that the provision allowing for the use of protected public land—if such use has minor impacts on the property and is approved by relevant resource agencies—has the potential to save time and has few challenges to implementation. State DOTs reported that the other SAFETEA-LU provisions GAO studied have both potential benefits and challenges but, in some cases, they identified alternative solutions that could better serve their needs. For example, although respondents indicated that they could save time by implementing the issue resolution process established in SAFETEA-LU, they also noted that the use of written agreements between highway project stakeholders—such as federal resource agencies—could better serve their purposes. Survey respondents also indicated that they are generally not interested in implementing two SAFETEA-LU provisions that would delegate environmental review decision-making authority from the Federal Highway Administration (FHWA) to states, primarily because the states did not want to accept federal court jurisdiction for the decisions made under those provisions.

States have implemented a variety of efforts to expedite highway projects and FHWA has initiated efforts to expedite projects by sharing innovative practices. For example, in 1997, the North Carolina DOT implemented a project development process that promotes early involvement of highway stakeholders and reduces permit processing times from years to months. Other state efforts are more recent, prompted by streamlining concepts promoted by FHWA beginning in 2010 under an effort known as Every Day Counts. Through Every Day Counts, FHWA encouraged states to consider implementing 15 specific innovative practices during 2011 and 2012, including 13 practices that could help expedite highway project completion. FHWA plans to introduce a new set of initiatives during 2012 for implementation during 2013 and 2014. FHWA developed performance measures for Every Day Counts and is currently collecting data to determine if these initiatives have had a positive impact on expediting highway projects.

ABBREVIATIONS

AASHTO	American Association of State Highway and Transportation Officials
CE	categorical exclusion
DOT	department of transportation
EA	environmental assessment
EIS	environmental impact statement
FHWA	Federal Highway Administration
FWS	U.S. Fish and Wildlife Service
MPO	metropolitan planning organization
NEPA	National Environmental Policy Act of 1969
NGO	nongovernmental organization
SAFETEA-LU	Safe, Accountable, Flexible, Efficient Transportation Equity Act: A Legacy for Users
USACE	U.S. Army Corps of Engineers

fulfills oversight functions. State and local governments execute the programs by matching and distributing federal funds; planning, selecting, and supervising projects; and complying with federal requirements.

Background Funding for highway projects represents a large federal investment—about $39 billion in fiscal year 2011. Federally funded highway projects are typically developed in the following four phases:

1. *Planning*. State DOTs and metropolitan planning organizations (MPO) assess the need for a project in relation to other potential highway project needs.
2. *Preliminary design and environmental review*. State DOTs identify potential transportation solutions based on needs identified during planning, potential environmental and social effects of those solutions, project cost, and construction location; analyze the effect, if any, of the proposed project and potential alternatives on the environment; and select the preferred alternative.
3. *Final design and right-of-way acquisition*. State DOTs finalize design plans, acquire property, and relocate residents and businesses.
4. *Construction*. State DOTs award construction contracts, oversee construction, and accept the completed project.

In the preliminary design and environmental review phase, many activities are carried out pursuant to the National Environmental Policy Act of 1969 (NEPA) and other federal laws. Under NEPA, federal agencies must assess the effects of major federal actions—those they propose to fund, carry out or permit—that significantly affect the environment. NEPA has two principal purposes: (1) to ensure that an agency carefully considers detailed information concerning significant environmental impacts and (2) to ensure that this information will be made available to the public. NEPA generally requires federal agencies to prepare documentation showing the extent of the project's environmental impacts. Per NEPA, the lead agencies—usually a state DOT and FHWA—will determine which of the three documentation types is needed as follows:

- Projects referred to as 'categorical exclusions' (CE) are determined to not individually or cumulatively have a significant effect on the quality of the environment. These projects require no or limited environmental review or documentation under NEPA. Examples of highway projects that are generally processed as CEs include resurfacing, constructing bicycle lanes, installing noise barriers, and landscaping. The vast majority of highway projects are processed as CEs (see figure 1). Based on data collected in 2009, FHWA estimates that approximately 96 percent of highway projects were processed as CEs.

- An environmental impact statement (EIS) is required for proposed projects that are determined to have a significant effect on the environment. In broad terms, FHWA starts the EIS process by publishing a notice of intent in the *Federal Register*. It then consults with resource agencies—such as USACE or FWS—and solicits comments from the public on a draft EIS, incorporates comments into a final EIS, and issues a record of decision. Among other things, the record of decision—which is the final step for agencies in the EIS process—identifies (1) the decision made; (2) the alternatives considered during the development of the EIS, including the environmentally preferred alternative; and (3) plans to mitigate environmental impacts. For the 32 projects in which FHWA was the lead agency and signed the EIS in fiscal year 2009, the average amount of time from signing the notice of intent to signing the record of decision was 83 months—almost 7 years.[1] As noted, FHWA estimates that based on its 2009 data approximately 1 percent of all federal-aid highway projects in the United States were processed with an EIS. While projects requiring an EIS are a small portion of all highway projects, they are likely to be high-profile, complex, and expensive projects. For these reasons, many efforts to expedite highway projects and reports which study those efforts, including this report, tend to focus on highway projects requiring an EIS.
- Project sponsors prepare an environmental assessment (EA) when it is not clear whether a project will have significant environmental impacts. An EA is intended to be a concise document that, among other things, briefly provides sufficient evidence and analysis for determining whether to prepare an EIS. If during the development of an EA, the project sponsor determines that the project will cause significant environmental impacts, the project sponsor will stop producing the EA and, instead, produce an EIS. However, an EA typically results in a finding of no significant impact, a document that presents the reasons why the agency has concluded that there are no significant environmental impacts to occur when the project is implemented. FHWA estimates that, based on its 2009 data, about 3

percent of all federal-aid highway projects were processed using an EA.

Numerous federal, state, and local laws determine the processes and tasks highway projects are to complete throughout the four phases. For example, SAFETEA-LU contains provisions that establish policies related to transportation planning and the environmental review process. Various environmental laws—including NEPA, the Endangered Species Act, the Clean Water Act, and the National Historic Preservation Act—establish processes and environmental requirements that projects must meet. Right-of-way acquisition must be accomplished according to the requirements of the Uniform Relocation Assistance and Real Property Acquisition Policies Act of 1970, as amended, a law designed to provide fair treatment of property owners and tenants when they are displaced by federally funded programs, including the construction of a federal-aid highway. Federal-aid highway projects are typically subject to a number of federally required contract provisions, such as nondiscrimination, payment of a predetermined minimum wage, and accident prevention. There are also numerous state and local laws—for example, several states, including California and North Carolina, have laws roughly equivalent to the federal NEPA—that projects must comply with and which help guide projects through various tasks in the process.

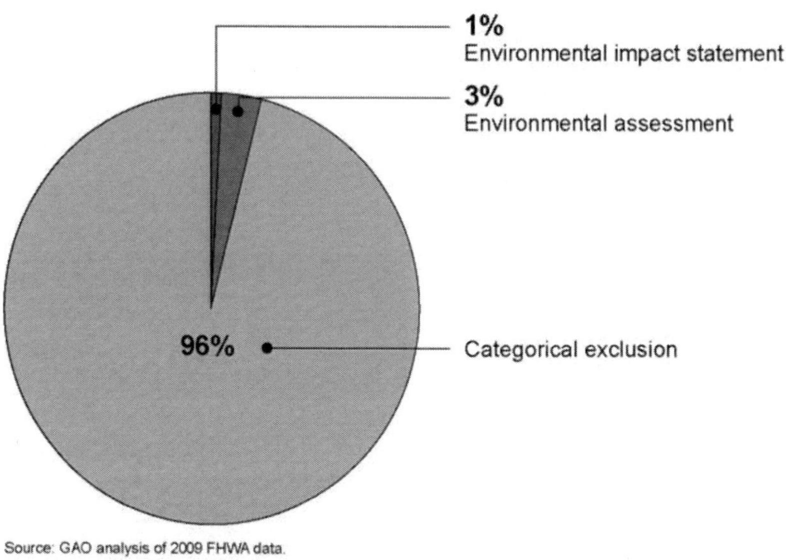

Figure 1. Federal Highway Projects by NEPA Action Class.

In addition, a number of provisions created by SAFETEA-LU are intended to help expedite highway projects. We analyzed seven of those provisions that have been implemented (see table 1), focusing primarily on those in Title VI of SAFETEA-LU, which deals with transportation planning and project delivery.[2] We surveyed officials from 52 state DOTs about the potential benefits and challenges associated with each of these SAFETEA-LU provisions and did not ask states to quantify these benefits or challenges. During survey pretesting, we learned that any number of variables could impact the time frames for completing a project, such as the SAFETEA-LU provisions we were asking about in our survey, the complexity of each highway project, or even the personalities of individuals working on tasks for the project. As such, our survey findings generally do not indicate specific values for the benefits and challenges (such as time savings) from implementing or using the SAFETEA-LU provisions, but rather represent state DOTs' perspectives (i.e., the degree to which they agree or disagree that a particular factor could be a benefit or a challenge) on the potential benefits and challenges of those provisions.

Table 1. SAFETEA-LU Provisions Meant to Help Expedite Highway Projects

GAO term[a]	SAFETEA-LU full title	Description	U.S. Code citation	SAFETEA-LU section
180-Day Statute of Limitations	Limitations on claims	Bars claims seeking judicial review of a permit, license, or approval issued by a federal agency for highway projects unless they are filed within 180 days after publication of a notice in the *Federal Register* announcing the final agency action, unless a shorter time is specified in the federal law under which the judicial review is allowed.	23 U.S.C. § 139(l)	6002
Categorical Exclusion Approval Authority	State assumption of response-bility for categorical exclusions	Authorizes U.S. DOT to assign and a state to assume responsibility for determining whether certain projects can be categorically excluded from the NEPA process.	23 U.S.C. § 326	6004
Design-Build Contracting[b]	Design build	Repealed the minimum cost requirements for use of design-build contracting for federal-aid highway projects.	23 U.S.C. § 112(b)(3)	1503(2)

Table 1. (Continued)

GAO term[a]	SAFETEA-LU full title	Description	U.S. Code citation	SAFETEA-LU section
		Also required the Secretary of Transportation to make changes to the design-build regulations, generally to permit a state transportation department to release requests for proposals and award design-build contracts prior to the completion of the NEPA process; however, it also precludes a contractor from proceeding with final design or construction before compl-etion of the NEPA process.		
Issue Resolution Process	Issue identification and resolution	Established procedures to resolve issues between state DOTs and relevant resource agencies.	23 U.S.C. § 139(h)	6002
Minor Impacts to Protected Public Land	Parks, recreation areas, wildlife and waterfowl refuges, and historic sites	Authorizes an historic site or publicly owned land from a park, recreation area, or wildlife or waterfowl refuge to be used for a transportation program or project if it is determined that such use would result in "de minimis impacts" to that resource.	23 U.S.C. § 138(b)	6009
NEPA Approval Authority	Surface transportation project delivery pilot program	Allows no more than five states to assume many federal environmental review responsibilities, in addition to determining whether certain projects can be categorically excluded from the NEPA process.	23 U.S.C. § 327	6005
Offering Financial Assistance to Stakeholder Agencies	Assistance to affected state and federal agencies	Allows a state to use its federal highway funds to support a federal or state agency participating in the environmental review process.	23 U.S.C. § 139(j)	6002

Source: Pub. L. No. 109-59.

[a] For purposes of this report, we have established terms for the SAFETEA-LU provisions we reviewed.

[b] Design-build contracting is a contracting method that combines the responsibilities for designing and constructing a project in a single contract instead of the more traditional approach of separating these responsibilities.

THE PROCESS TO COMPLETE HIGHWAY PROJECTS IS COMPLEX AND LENGTHY DUE TO MULTIPLE FACTORS

Completing a highway project can involve many stakeholders—including federal, state, and local government agencies; nongovernmental organizations (NGO); and private citizens—and, for major highway projects, as many as 200 steps from planning through construction (see figure 2). A number of additional factors can also affect project time frames.

Wide Range of Stakeholders

A wide range of stakeholders can be involved in highway projects, from federal, state, and local agencies with varying missions and responsibilities to NGOs, contractors, and private citizens. Different factors, however, will help determine the extent to which stakeholders will become involved in the project. For example, if a highway project will not affect endangered or threatened species, it is likely that FWS—which is responsible for implementing the Endangered Species Act for freshwater and terrestrial species—will not become involved in the project. Additionally, some states have developed written agreements—known by a number of terms, including programmatic agreements or memoranda of agreement—with other state or federal agencies that can help to establish a process for consultation, review, and compliance with one or more federal laws, allowing for the project to be reviewed more quickly. Regardless, there are a host of stakeholders that could become involved in a highway project as follows:

- *Transportation agencies*. Federal and state transportation agencies are responsible for improving, maintaining, and planning highway systems with a focus on safety, reliability, effectiveness, and sustainability. Among other things, FHWA oversees planning and project completion by reviewing statewide long-range transportation plans, evaluating whether a project meets environmental protection requirements, and authorizing acquisition of property for highway projects it funds. State DOTs are typically the focal point for project planning and construction and are responsible for setting the relevant goals for the state, planning safe and efficient transportation, designing most projects, identifying and mitigating environmental impacts, acquiring property for highway projects, and awarding and overseeing construction contracts.

- *Federal resource agencies.* Federal resource agencies, such as those described below, are responsible for managing and protecting natural and cultural resources like wetlands, historic properties, forests, and wildlife:
  - *The Advisory Council on Historic Preservation*, established by the National Historic Preservation Act of 1966, seeks to promote the preservation, enhancement, and sustainable use of the nation's historic resources. The council advises the President and the Congress on national historic preservation policies and ensures federal agencies take such issues into account when developing and implementing federal projects.
  - *USACE* issues permits for the dredging and filling of waters of the United States, including wetlands within the agency's jurisdiction, under Section 404 of the Clean Water Act.
  - *The Environmental Protection Agency* administers, among other things, the Clean Air and Clean Water Acts.
  - *FWS* implements the Endangered Species Act with respect to freshwater and terrestrial species.
  - *The National Marine Fisheries Service* implements, among other things, the Marine Mammal Protection Act and the Endangered Species Act with respect to most marine species and anadromous fishes (which spend portions of their life cycle in both fresh and salt water).
  - *The U.S. Forest Service* transfers land for highway rights of way within the National Forest System to states through FHWA.
- *State resource agencies.* These state-level agencies are generally responsible for managing and protecting the state's natural and cultural resources.
  - State resource agencies, like their federal counterparts, participate in and review assessments of environmental impacts, in accordance with their responsibilities under federal or state laws.
  - A state historic preservation office advises and consults with federal and other state agencies to identify historic properties and assess and resolve adverse effects to them under the National Historic Preservation Act.
- *Local governments.* Local governments involved in highway projects include MPOs and rural planning organizations.

	Planning	Preliminary design and environmental review	Final design and right-of-way acquisition	Construction
Potential stake-holders involved	State DOTs and Federal Highway Administration generally involved in all aspects of highway projects			
	• Local governments (such as metropolitan planning organizations or rural planning organizations)	• Federal resource agencies (such as U.S. Army Corps of Engineers for projects impacting wetlands, or U.S. Fish and Wildlife Service for projects impacting threatened and endangered plant and animal species) • State resource agencies (such as the state historic preservation office for projects impacting historic property) • Local governments • Nongovernmental organizations • Contractors • Private citizens	• Federal resource agencies • State resource agencies • Contractors • Private citizens	• State resource agencies • Contractors
Typical steps	• Assess transportation purpose and need • Solicit public comment • Gain approval to be included in the state's 20 year plan • Gain approval to be included in the state's short-term program (at least 4 years) for projects that are to be implemented, with expectation that funds will be available • Determine sources of funding	• Consider alignment issues and required lanes • Identify alternatives, including not building the project, to minimize potential harm to the environment and historic sites • Select preferred alternative • Prepare a preliminary design of the highway • Solicit comments on the project and its potential effects from private citizens and from local governments • Gain concurrence from federal agencies from which environmental and historic preservation concurrence is required	• Finalize design plans • Appraise and acquire property • Relocate utilities and affected citizens before construction, if necessary • Finalize project cost estimates	• Advertise and evaluate bids; award contracts • Begin construction • Resolve unexpected problems • Accept delivery

Source: GAO.

Figure 2. Potential Stakeholders and Typical Steps Involved in a Major New Highway Project.

- Every urbanized area with a population of 50,000 or more has an MPO, an organization made up of representatives of local governments—county, city, and town government officials—for the purpose of transportation planning and coordination of highway and transit projects. According to a nonprofit organization that represents MPOs, there are almost 400 MPOs in the United States.
- Rural planning organizations are typically voluntary planning organizations that serve as a forum for local officials to develop consensus on regional transportation priorities for an area with a population of less than 50,000.

- *NGOs*. NGOs advocate for a number of issues, including the environment and transportation. Examples of NGOs include the following:

- The Natural Resources Defense Council is an environmental organization that seeks to protect the environment by educating the public, lobbying government officials, and litigating, if necessary.
- AASHTO advocates for transportation-related policies and provides technical transportation-related support to states.

- *Contractors.* Contractors generally are private sector companies that bid on contracts from federal and state transportation agencies to conduct various activities, such as conducting environmental studies or constructing a highway.
- *Private citizens.* Private citizens have the opportunity to provide comments and opinions in venues like public hearings.

Large Number of Steps in the Process

In addition to the involvement of a large number of stakeholders, completing a major highway project takes a number of years because of the many tasks, requirements, and approvals involved throughout the four phases of a highway project. Major highway projects can involve as many as 200 steps from the initial planning phase through the construction phase that require actions, approvals, or input from a number of stakeholders.[3]

Planning Phase

State DOTs and local planning organizations assess a project's purpose and consider the need for the project in relation to the need for other potential highway projects. To receive federal transportation funding, any project in an urbanized area must emerge from the relevant MPO and state DOT planning processes. For nonmetropolitan areas not covered by an MPO, states must consult with and provide opportunities for local officials to participate in statewide planning. To meet federal planning requirements, states must develop (1) a long-range statewide transportation plan covering a 20-year period and (2) a state transportation improvement program—that is, the state program of transportation projects covering at least a 4-year period that are to be supported with federal surface transportation funds, as well as regionally significant projects requiring an action by FHWA, whether or not federally funded.

Preliminary Design and Environmental Review Phase

During preliminary design, a project's location and design are identified, along with the effect, if any, of the proposed project and of potential alternatives on the environment; eventually, a preferred alternative is selected. Among other tasks, state DOTs identify the preliminary engineering issues, proposed alignment of roadways, and costs, as well as create topographic surveys and conduct traffic studies. During environmental review, the proposed project alternatives are examined and may require review, input, or feedback from relevant resource agencies such as USACE, FWS, or the Environmental Protection Agency. Environmental reviews require state and FHWA officials to address and comply with many federal laws—FHWA has identified over 40 environmental laws—as well as applicable state laws. More complex projects require additional time for the completion of preliminary designs and environmental reviews. In addition, private citizens and local governments are asked to comment on the project and its potential effects. At the end of this phase, the preferred alternative is selected.

Final Design and Right-of-Way Acquisition Phase

State DOTs finalize design plans, acquire property, and relocate utilities in the final design and right-of-way acquisition phase. State DOTs develop detailed engineering plans consistent with environmental documents and updated environmental studies, and finalize cost estimates. If a significant amount of time has passed since the preliminary design work was performed, right-of-way maps and other information may need to be updated. Acquiring property for the project includes determining any restrictions to state ownership of the property, determining the identities of property owners, making offers to property owners based on just compensation, negotiating a purchase price, relocating property owners and tenants, and sometimes invoking eminent domain. Utilities must be located, marked, surveyed, and possibly relocated. If there are a significant number of underground utilities, professional engineers, geologists, and land surveyors may be needed to determine the exact location of the utilities.

Construction Phase

State DOTs award construction contracts, oversee construction, and accept the completed project. State DOTs request and evaluate bids on projects and then award the contract. The federal government is not directly involved in construction, but does have an oversight role. For example, projects that receive federal-aid highway funds require FHWA concurrence on the award. During construction, the contractor and the state resolve any unexpected problems that may arise, such as removal of hazardous waste at the construction site. Once satisfied that construction has been carried out as agreed to with the contractor, the state must approve the final completion of construction.

Additional Factors That Can Affect Time Frames

In addition to the many stakeholders and tasks involved, a number of other factors can complicate the process and lead to longer highway project time frames such as the following:

Funding Availability

The availability of funding for large highway projects can affect how long it takes to complete a project. For example, one state DOT informed us it has completed a number of EISs for highway projects, but that these projects are stalled due to a lack of funds. In addition, a state DOT official stated that since the state did not have enough funding to complete major highway projects, they are choosing to focus more on completing smaller, less expensive highway projects such as bridge replacements and repaving. Of those responding to our survey, most state DOTs identified funding as a challenge for all project phases but found it to be more of a challenge in both the planning phase and the preliminary design and environmental review phase.[4]

Changing Transportation Priorities

Changes in a state's transportation priorities during a project's duration can complicate time frames and delay the project. For example, one administration may favor a highway project when it is first planned and may provide the necessary financial support; however, a new administration with different priorities may come in before the project is completed and withdraw or reduce support and funding. If a project that was shelved garners support again, in some cases, FHWA, the state DOT, or resource agencies might have to reevaluate, rework, and update environmental- or NEPA-related documents and information to ensure that the environmental impact information is current. This can lead to a longer project time frame.

Project Opposition

Public opposition and litigation can also lengthen highway project time frames or even lead to the cancellation of a project. For example, the Elizabeth Brady Road project in Orange County, North Carolina was canceled by FHWA due to public and local government opposition to the project. After the project began the preliminary design and environmental review phase, local community and government officials determined that there was insufficient need for the project because the potential costs outweighed the project's potential benefits. As a result, local government officials withdrew their support for the project and it was canceled. Public controversy related to a highway project can sometimes lead to litigation, which can also lengthen highway project time frames. Litigants might settle their lawsuit if, for example, a state DOT agrees to change the design of a project to limit its impact on a species or increase noise abatement measures. Lawsuits can also lead to longer completion time frames. For example, plaintiffs filed suit in 2006 against FHWA and the U.S. Forest Service for a highway project in Alaska, alleging that these parties failed to comply with a number of federal laws, including NEPA. The U.S. District Court found that the final EIS issued for the project was not valid and issued an injunction stopping all work on the project. Upon appeal, the U.S. Court of Appeals for the Ninth Circuit upheld the District Court decision.[5] In September 2011, nearly 5 years after the lawsuit was filed, the Alaska Department of Transportation and Public Facilities began work to prepare a supplemental EIS—that is, an updated EIS—for the project. The agency anticipates issuing a record of decision for this project in late 2013.

State DOTs Generally Agree That SAFETEA-LU Provisions Could Decrease Project Time Frames but Find Some Provisions More Useful Than others

States identified both benefits and challenges with each of the SAFETEA-LU provisions meant to help expedite highway projects but acknowledged alternative solutions for some of the provisions that better served their purposes. In our survey, state DOTs most frequently agreed that the Minor Impacts to Protected Public Land provision of SAFETEA-LU has the potential to save time (see table 2) and has relatively few challenges to implementation.[6]

Table 2. Potential Time Savings and other Key Findings States Reported Regarding SAFETEA-LU Provisions Meant to Help Expedite Highway Projects, since 2005

SAFETEA-LU provision	State agreement on potential time savings	Key findings states reported in GAO survey
Minor Impacts to Protected Public Land	92% (47 of 51 states)	States generally did not identify potential challenges with this provision. 82% of states disagreed that participation requirements for this provision are too challenging to fulfill, and nearly all states responding (47 of 49) have used this provision at least once.
Design-Build Contracting	79% (30 of 38 states)	Smaller highway projects, which comprise the majority of all highway projects, generally do not lend themselves to design-build contracting. Additionally, some states are prohibited by state statute from using this contracting method.
180-Day Statute of Limitations	78% (32 of 41 states)	Some states expressed concern that a shorter statute of limitations could draw undue attention to the project and encourage litigation.
Offering Financial Assistance to Stakeholder Agencies	77% (34 of 44 states)	Some states indicated that use of this provision has created a better working relationship between highway stakeholders; others noted that they saw limited results and have since stopped providing funding.
Categorical Exclusion Approval Authority	76% (34 of 45 states)	Almost two-thirds of states (29 out of 49) agreed that programmatic agreements could serve their agency better than this initiative. Only three states—Alaska, California, and Utah—participate in this program.
Issue Resolution Process	61% (22 of 36 states)	No state has used this provision, choosing instead to resolve issues at the lowest possible staff level or to follow procedures established in programmatic agreements.
NEPA Approval Authority	56% (19 of 34 states)	States expressed reluctance to accept federal court jurisdiction in order to participate in this pilot program, an action they generally refer to as 'waiving their sovereign immunity.' Only one state—California—is participating in this pilot program.

Source: GAO.

Minor Impacts to Protected Public Land

Most respondents agreed that the Minor Impacts to Protected Public Land provision of SAFETEA-LU has potential time savings benefits, and nearly all have used this provision at least once. This provision authorizes an historic site or publicly owned land from a park, recreation area, or wildlife or waterfowl refuge, to be used for a transportation program or project if a DOT determines that such use would result in minor impacts (i.e., "de minimis impacts") to that resource.[7] The Department of Transportation Act of 1966 includes a provision—known as Section 4(f)—which stipulates that FHWA and other DOT agencies cannot approve the use of land from publicly owned parks, recreational areas, wildlife and waterfowl refuges, or public and private historical sites unless (1) there is no feasible and prudent alternative to the use of such land and (2) the action includes all possible planning to minimize harm to the property resulting from use.[8] Complying with Section 4(f) can result in additional time to receive project approval. One NGO we spoke with noted that use of the Minor Impacts to Protected Public Land provision of SAFETEA-LU is a more "common sense" approach that not only allows greater use of these protected properties when only very minor impacts are likely to occur, but also helps to expedite highway projects.

- *Potential benefits*. Of those responding to our survey, 92 percent of states (47 of 51 states) agreed that this SAFETEA-LU provision has the potential to save time. In addition, of those responding, 80 percent (41 of 51 states) identified the Minor Impacts to Protected Public Land provision as having the potential to create staffing or personnel savings and 59 percent (29 of 49 states) identified the provision as having the potential to increase the number of projects completed.
- *Potential challenges*. Most states who responded to our survey did not indicate significant challenges to implementing this SAFETEA-LU provision. For example, 82 percent of states (42 of 51 states) disagreed that the participation requirements for this provision are too challenging to fulfill, indicating that this provision may be easier to use or implement than the other provisions.
- *Implementation/use*. Of all the SAFETEA-LU provisions we studied, the Minor Impacts to Protected Public Land provision was used most frequently. Of those states responding, almost all (47 of 49 states) had used this provision at least once, with 9 states indicating that they have used this provision for more than 50 percent of their highway projects since SAFETEA-LU's enactment in 2005.

Design-Build Contracting

Most states responding to our survey agreed that the Design-Build Contracting provision within SAFETEA-LU has the potential to save time, but many states have not used this contracting method and, therefore, have not had the opportunity to take advantage of this provision. Under the traditional procurement approach, design and construction services must be separated and a construction contract, which generally goes to the lowest bidder, can be awarded only after the design is complete. Design-build contracting combines the responsibilities for designing and constructing a project in a single contract instead of separating these responsibilities. The Design-Build Contracting provision in SAFETEA-LU repealed the minimum cost requirements for use of design-build contracting for federal-aid highway projects; prior to enactment of SAFETEA-LU, federal-aid highway projects needed to have total costs exceed $50 million in order to use design-build contracting.

In our survey, state DOTs generally agreed that the Design-Build Contracting provision has the potential to save time, but noted some challenges and limited use.

- *Potential benefits*. Of those responding, 79 percent of states (30 of 38 states) agreed that this SAFETEA-LU provision has the potential to save time. Fewer states that responded agreed that other benefits could potentially be realized from use of design-build contracting: 45 percent (17 of 38 states) noted that its use could potentially increase the number of highway projects completed, and 37 percent (14 of 38 states) noted potential staff or personnel savings.
- *Potential challenges*. Most states did not indicate significant challenges to using this SAFETEA-LU provision in the survey questions we asked.[9] However, states did provide some challenges to design-build contracting in their written responses. For example, some states are prohibited by state statute from using design-build contracting for highway projects. Other states noted that problems in completing other project tasks, such as obtaining permits, can slow overall project completion time frames such that potential time savings achieved by design-build contracting might be negated.
- *Implementation/use*. Of those responding, 60 percent of states (26 of 43 states) have used design-build contracting at least once since enactment of SAFETEA-LU. However, the majority of states that responded (24 of 43 states, or 56 percent) use design-build contracting for less than 10 percent of all highway projects. States noted both in

our survey and in our interviews that smaller highway projects—such as resurfacing or landscaping projects that are processed as CEs—generally do not require extensive design work and, as a result, do not lend themselves to the use of design-build contracting.

180-Day Statute of Limitations

Most states responding to our survey agreed that the 180-Day Statute of Limitations provision has potential benefits, and many have had at least one highway project since SAFETEA-LU's enactment that has taken advantage of it. Prior to enactment of SAFETEA-LU, individuals or organizations generally had up to 6 years in which they could file a judicial claim on a final agency action related to environmental requirements, such as NEPA requirements. This provision of SAFETEA-LU bars claims seeking judicial review of a permit, license, or approval issued by a federal agency for a highway project unless that claim is filed within 180 days of a notice in the *Federal Register*—FHWA generally publishes these notices—announcing the final agency action.

In our survey, state DOTs generally agreed that the 180-Day Statute of Limitations provision has the potential to save time, and many states have taken advantage of this provision since SAFETEA-LU's enactment; however, some states expressed concerns that a shorter statute of limitations could actually encourage litigation.

- *Potential benefits.* Of those responding, 78 percent (32 of 41 states) agreed that this SAFETEA-LU provision has the potential to save time. 56 percent of those states responding (22 of 39 states) also agreed that this provision could result in staff or personnel savings. Only about one-third of those responding (15 of 41 states, or 37 percent) agreed that the provision could result in more projects being completed.
- *Potential challenges.* When asked what challenges, if any, could be faced from this SAFETEA-LU provision, 8 states noted that a shorter statute of limitations may actually encourage litigation.[10] In general, these 8 states noted that if the shorter statute of limitations was used, such use could be seen as suspect by outside entities and encourage them to question the project and file a lawsuit against it.
- *Implementation/use.* Due in part to the above mentioned challenge, at least one state has chosen not to take advantage of the shorter statute of limitations. Of those states responding to our survey, 64 percent (29 of 45 states) have used the 180-Day Statute of Limitations provision

for at least one project since enactment of SAFETEA-LU, leaving 36 percent of states (16 of 45 states) as having never used the provision. Officials from one state DOT we interviewed did note that they have chosen to not pursue this shorter statute of limitations as they feel its use might draw undue attention to the project and encourage outside entities to litigate it.

Offering Financial Assistance to Stakeholder Agencies

Most states responding to our survey agreed that the Offering Financial Assistance to Stakeholder Agencies provision of SAFETEA-LU has potential benefits, including time and staffing or personnel savings, but fewer states have actually taken advantage of this provision. Under this SAFETEA-LU provision, a state DOT can use part of its federal highway funding to support staff for a federal or state agency participating in the environmental review process, such as the local USACE or FWS office. Funds provided in accordance with this provision may only be used for projects in a given state that support activities that directly and meaningfully contribute to expediting and improving transportation project planning and completion.

In our survey, state DOTs generally agreed that the Offering Financial Assistance to Stakeholder Agencies provision has the potential to save time, but its use is not as widespread as some of the other SAFETEA-LU provisions.

- *Potential benefits.* Of those responding, 77 percent of states (34 of 44 states) agreed that this SAFETEA-LU provision has the potential to save time. The majority of those states responding also agreed that this provision could have potential staff or personnel savings (26 of 44 states, or 59 percent), as well as increase the number of projects completed (25 of 44 states, or 57 percent).
- *Potential challenges.* States responding to our survey generally noted some challenges to using this SAFETEA-LU provision. Nineteen of 44 states (43 percent) responding agreed that programmatic agreements could serve their agency better than this SAFETEA-LU provision. However, only 9 percent of states (4 of 43 states) agreed that a state or agency policy would discourage them from providing financial assistance to affected entities.
- *Implementation/use.* Of those responding, 58 percent of states (25 of 43 states) have provided financial assistance to affected entities at least once. However, a large number (18 of 43 states, or 42 percent) have never taken advantage of this provision. In our interviews with

state DOTs and federal resource agencies, interviewees also had mixed opinions on this SAFETEA-LU provision. For example, some interviewees stated that use of this SAFETEA-LU provision has created a better working relationship between the state DOT and the affected entity. However, other states we interviewed indicated that they had previously provided financial assistance to affected entities but had seen limited results and had stopped providing such funding. Staff from the federal resource agencies we spoke with were generally familiar with this SAFETEA-LU provision and, in some cases, found it to be helpful in expediting the completion of highway projects.

Categorical Exclusion Approval Authority

While most states responding to our survey agreed that the Categorical Exclusion Approval Authority provision within SAFETEA-LU has the potential to save time, only three states are participating in this program, and most states indicated that other techniques could achieve the same outcome as this program. This SAFETEA-LU provision authorizes U.S. DOT to assign and a state to assume responsibility for determining whether certain designated activities constitute actions that are categorically excluded from the requirement to prepare an EA or EIS. As noted above, most highway projects in the United States are processed as CEs, thus many of the projects a state DOT leads could be affected by participation in this program. As of April 2012, only three states are participating in this program: Alaska, California, and Utah. These three states have signed memoranda of agreement with their respective FHWA division offices outlining the processes and procedures they are to follow once assuming authority to approve CEs. Per SAFETEA-LU, these agreements are to last no more than 3 years, but can be renewed by mutual agreement of both the state DOT and FHWA. States that choose to participate in this program are required to accept federal court jurisdiction for the decisions they make under the program.[11] Highway stakeholders often refer to this aspect of the Categorical Exclusion Approval Authority provision as requiring the state legislature to 'waive its sovereign immunity.'

In our survey, state DOTs generally agreed that the Categorical Exclusion Approval Authority provision has the potential to save time, but several respondents supported the use of approaches other than this program to achieve a similar outcome.

- *Potential benefits.* Of those responding, 76 percent (34 of 45 states) agreed that this SAFETEA-LU provision has the potential to save time. States also saw this provision as having the potential to increase the number of projects being completed (26 of 44 states, or 59 percent) and create staffing or personnel savings (22 of 45 states, or 49 percent).

- *Potential challenges.* The majority of those responding to our survey (29 of 49 states, or 59 percent), as well as some state DOTs we spoke with, indicated that the use of agreements—such as programmatic agreements or memoranda of agreement—could serve the state DOTs better than this SAFETEA-LU provision. Seventeen state DOTs noted in our survey that they have undertaken efforts to establish agreements with their respective FHWA division offices or federal and state resource agencies. Among other things, these agreements establish policies and procedures for the state DOTs to follow in certain situations and scenarios. For example, the Missouri DOT has entered into a programmatic agreement with the FHWA division office to allow the state DOT to classify certain activities specified in the agreement as CEs without submitting each project to FHWA for approval of an environmental classification of CE.
  Agreements such as these allow relevant agencies—in this case, the FHWA division office—to make certain that projects comply with relevant laws and regulations but relieve the agency of the burden of having to review every project that the state DOT undertakes.
- *Implementation/use.* As noted above, only three states—Alaska, California, and Utah—are participating in the program created by this SAFETEA-LU provision. All three state DOTs indicated that they have seen positive outcomes from their participation in the program.

Issue Resolution Process

States saw the Issue Resolution Process provision within SAFETEA-LU as having some potential to save time, but none has used this provision, and most saw the use of written agreements between parties—including programmatic agreements or memoranda of understanding—as a better alternative. This SAFETEA-LU provision established procedures for resolving issues that could delay completion of the environmental review process or could result in denial of approvals required for the project under specific laws, such as the Clean Water Act or the Endangered Species Act. In general terms, a meeting of the relevant agencies can be convened to resolve the issues at hand; if a resolution cannot be achieved, the lead agency—for most federal-aid highway projects, this would be FHWA—is to notify a number of interested parties, including the Senate Committee on Environment and Public Works, the House Committee on Transportation and Infrastructure, and the Council on Environmental Quality within the Executive Office of the President.

In our survey, state DOTs indicated that this SAFETEA-LU provision has some potential to create time savings but generally saw the use of programmatic agreements as a better alternative for resolving issues between parties.

- *Potential benefits.* Of those responding, 61 percent (22 of 36 states) agreed that this SAFETEA-LU provision has the potential to save time. States generally did not agree that other potential benefits could arise from the use of this SAFETEA-LU provision: 37 percent (14 of 38 states) agreed that its use has the potential to create staffing or personnel savings, and only 29 percent (11 of 38 states) agreed that its use could increase the number of projects completed.
- *Potential challenges.* The majority of the states responding to this portion of the survey (25 of 41 states, or 61 percent) indicated that established agreements, like a programmatic agreement, could better serve their agency than this SAFETEA-LU provision. Some of the state DOTs we interviewed indicated that they had programmatic agreements in place with various parties, such as federal resource agencies, that established procedures by which issues could be identified and resolved. States and federal resource agencies told us that they would prefer if issues were identified and resolved at lower staff levels, rather than by management or executives, or through the process established in this SAFETEA-LU provision.
- *Implementation/use.* As noted above, this SAFETEA-LU provision has not been used or implemented, and highway stakeholders we interviewed noted that resolving these disputes using methods other than this SAFETEA-LU provision are preferred.

NEPA Approval Authority

The majority of states responding to our survey agreed that the NEPA Approval Authority provision within SAFETEA-LU has the potential to save time, but most states indicated that it is too burdensome to begin participating. This SAFETEA-LU provision required the establishment of a pilot program to permit not more than five states to assume certain federal environmental review responsibilities, such as the environmental reviews required under NEPA or other federal laws.[12] SAFETEA-LU listed five states that were given the opportunity to participate in this pilot program: Alaska, California, Ohio, Oklahoma, and Texas. To date, California is the only state that is participating in this pilot program. Other states expressed interest but withdrew their applications to participate. Eventually, FHWA opened the pilot program to all states, but limited participation to a total of five states, as called for in SAFETEA-LU. Much like the Categorical Exclusion Approval Authority provision of SAFETEA-LU, states that choose to participate in this program are required to accept federal court jurisdiction for the decisions they make under the program, an action which is generally undertaken by the state legislature and which highway stakeholders often referred to as requiring the state legislature to 'waive its sovereign immunity.'

In our survey, state DOTs agreed that the NEPA Approval Authority pilot program has the potential to save time, but a majority of respondents indicated that participation requirements for this provision are too challenging to fulfill.

- *Potential benefits.* Of those responding, 56 percent of states (19 of 34 states) agreed that this SAFETEA-LU provision has the potential to save time. States generally agreed that this provision does not have the potential to save staffing or personnel resources, or increase the number of projects completed.[13]
- *Potential challenges.* The majority of states responding to this section of our survey (27 of 33 states, or 82 percent) indicated that the participation requirements for this initiative are too challenging to fulfill. This message was reiterated in interviews we conducted with state DOTs. For example, officials from these agencies stated that accepting federal court jurisdiction for the environmental review decisions they make was something they, their agency management, or their state legislature—which would need to approve the acceptance of such responsibility—did not wish to take on.
- *Implementation/use.* As noted above, California is the only state that is currently participating in this pilot program. According to the California Department of Transportation, highway projects requiring an EA now take about 30 months less to complete than they previously did. In addition, staff from some of the federal resource agencies we spoke with indicated that California's participation in the pilot program has generally been beneficial, with staff from one resource agency calling for California's continued participation in the pilot program. While California has reported a time savings from its participation in the NEPA Approval Authority pilot program, other

states with whom we spoke did not express interest in this pilot program, with most states citing the requirement to accept federal court jurisdiction for the decisions they make under the program as a key reason why they do not wish to participate. In addition, at least two states indicated that they appreciate having FHWA make these environmental decisions. More specifically, they stated that FHWA has the staff and expertise to make informed decisions regarding environmental impacts.

STATES AND FHWA HAVE INITIATED EFFORTS TO DEVELOP AND SHARE INNOVATIVE PRACTICES FOR EXPEDITING HIGHWAY PROJECTS

States have implemented a variety of efforts to expedite highway projects, and FHWA has initiated efforts to share innovative practices. Some state efforts began in the 1990s in response to challenges faced at that time. Other state efforts are more recent, prompted by new authorities provided by SAFETEA-LU or by streamlining concepts recently promoted by FHWA. FHWA is making efforts to share innovative practices to help expedite highway projects, most recently through an effort known as Every Day Counts. However, it is too soon to determine the effect these initiatives have had on highway project time frames.

States Have Implemented Streamlining Practices Spanning All Project Phases

Most states have made efforts to expedite projects with state DOTs playing a key role in choosing the techniques that are used. In our survey of state DOTs, we asked officials about initiatives they have undertaken since the enactment of SAFETEA-LU to expedite the four phases of highway projects. Most states—43—reported that they have implemented at least 1 initiative, 4 states reported undertaking no initiatives, 3 states did not respond for any phase, and 2 states reported no initiatives for some phases and no response for other phases. According to the survey, states most often implemented initiatives involving the preliminary design and environmental review phase (39 states). Twenty-two states reported implementing initiatives involving the planning phase, 15 states involving the final design and right-of-way acquisition phase, and 19 states involving the construction phase. We also asked officials about the potential benefits that could be realized from the initiatives they had undertaken. For each of the four phases of a highway project, time savings was the benefit most often cited by states.[14] Staff savings was cited as a potential benefit by a majority of states for all phases except construction, when it was cited as a potential benefit by 39 percent of the states (7 of 18 states) responding. Increased number of projects completed was cited as a benefit by a majority of officials responding for all phases except construction, where it was cited as a potential benefit by half of the states responding (9 of 18).

State DOTs reported implementing a variety of types of initiatives to expedite highway projects but generally not one type more than another. In fact, only 4 of more than 30 initiatives were reported by 10 or more states:

- *Linking Planning and Environmental Review.* Twenty-three states reported implementing steps that linked their planning and environmental review processes. Using information collected in the planning phase and carrying it through the environmental review phase can minimize duplication of effort and reduce delays in project implementation. For example, the North Carolina Department of Transportation designed a project development process, implemented in 1997, that promotes early involvement of state and federal stakeholders. Each project must pass seven concurrence points that cover aspects of project planning, environmental review, and permitting. This process reduces permit processing times from years to months, according to North Carolina Department of Transportation officials.
- *Using Programmatic Agreements.* Seventeen states reported implementing programmatic agreements. These written documents establish a process for consultation, review, and compliance with one or more federal laws between one or more parties, such as a state DOT and a resource agency. Programmatic agreements can help reduce project time frames. For example, an agreement between the Illinois Department of Transportation and FHWA created both a procedure for negotiating project-specific time frames for completing environmental reviews and completion time goals for EISs and EAs. After processing five EISs and four EAs under the agreement, project completion time was reduced by at least 2 years, according to a 2010

AASHTO report for FHWA. Some state DOTs have used programmatic agreements for more than a decade, including at least four states that have used programmatic agreements since the 1990s. An agreement between the California Department of Transportation, FHWA, and four resource agencies has been in place since 1991.

- *Using Design-Build Contracts.* Eleven states reported implementing design-build contracts. Again, design-build contracting combines the responsibilities for designing and constructing a project in a single contract instead of separating these responsibilities. Design-build contracting can provide significant time savings compared with the design-bid-build approach where design and construction phases must take place in sequence, according to FHWA.
- *Using Other Nontraditional Construction Contracts.* Eleven states reported implementing other nontraditional construction contracts. These included construction manager/general contractor contracts[15] and other nontraditional contract approaches such as cost plus time bidding, lane rentals, and contractor completion incentives and disincentives.[16]

State DOTs also reported implementing several other highway project streamlining initiatives, including use of electronic bidding, clarifying the scope of preliminary design, linking the final design and right-of-way acquisition phase with prior project development phases, as well as early right-of-way purchases.

FHWA Is Sharing Practices among States Though It Is Too Soon to Know the Outcome

FHWA is sharing information on methods to expedite highway projects with state DOTs through an effort called Every Day Counts. This effort's goals are to shorten project time frames and accelerate use of technology and innovation by convincing states to adopt proven, rapidly deployable innovations. Many of these innovations were in use by some states before they were selected for promotion through Every Day Counts: as discussed earlier, for example, California's 1991 programmatic agreement or North Carolina's 1997 project development process. FHWA selected its Every Day Counts innovations through a process that involved headquarters and division office staff, as well as outside organizations such as AASHTO, Associated General

Contractors, and the American Road and Transportation Builders Association. Every Day Counts was introduced at AASHTO meetings in spring 2010 and subsequently promoted at 10 regional summits sponsored by FHWA and AASHTO. Each state was asked to decide on specific initiatives it wanted to pursue and develop a plan for implementing them during 2011 and 2012. States were also asked to create transportation innovation councils to track attainment of goals. Under Every Day Counts, FHWA urged state DOTs to consider use of 15 specific initiatives—10 designed to shorten project time frames and 5 designed to accelerate technology deployment. The Every Day Counts initiatives described below include 3 of the 4 initiatives that states have taken as described above. See table 3 for brief descriptions of the initiatives in FHWA's Every Day Counts effort for implementation in 2011 and 2012 and appendix II for more detailed information on those Every Day Counts initiatives. FHWA expects to introduce a new set of initiatives in late 2012, to be implemented during 2013 and 2014, and intends that another series of initiatives will follow for the period from 2015 to 2016. FHWA began, in October 2011, the process of soliciting ideas for new initiatives to implement in 2013, saying it would consider proposed initiatives based on factors such as transportation system impact, readiness for deployment, affordability, and urgency.

Table 3. FHWA's Every Day Counts Initiatives for 2011 and 2012

Title	Description
Initiatives to shorten project time frames	
Linking planning and environmental review	Promotes use of planning documents and decisions from the project planning process in the environmental review process. This can minimize duplication of effort and reduce delays in project implementation.
Early consultation with FHWA environmental attorneys	Consulting with FHWA environmental attorneys at early decision points can help decision makers avoid problems later, saving time and costs.
Expanded use of programmatic agreements	A programmatic agreement formally spells out the terms of an agreement between a state DOT and other state and/or federal agencies.
Compensatory mitigation	Various laws require compensatory mitigation for projects that cause unavoidable impacts to streams, wetlands, and other waters. Compensatory mitigation can be accomplished through mitigation banks or in-lieu fee programs.[a]
Clarifying scope of preliminary design	States can complete design activities not required for a NEPA determination under preliminary design. When performed concurrently with the NEPA process, these activities can expedite highway projects without affecting eligibility for federal aid.

Title	Description
Allowable right-of-way acquisition streamlining	To save time, states can acquire property using techniques such as appraisal waiver valuations, incentive payments to advance acquisition and relocation, and appraisals and negotiations of property acquisition (up to $10,000) by the same individual.
Effective coordination for utility relocation	States can effectively coordinate utility relocation during project development by techniques such as documenting the terms and considerations for accomplishing utility relocation, and financing the work in an effective and timely manner.
FHWA technical assistance for stalled projects	FHWA subject matter experts, drawn from offices nationwide, provide assistance to highway stakeholders to help resolve issues that would otherwise hold up the NEPA review. These staff can help resolve resource-specific issues concerning wetlands, endangered species, and cultural resources.
Use of design-build contracts	Design-build contracts combine design and construction tasks into one contract, eliminating the separate bid phase and allowing certain aspects of design and construction to take place at the same time.
Use of construction manager/general contractor contracts	The owners of a project hire either a general contractor or design firm to serve as the construction manager, placing responsibility for design review, design modifications, system integration, and construction with that single contractor.
Initiatives to accelerate technology deployment	
Warm mix asphalt paving	Asphalt is produced and placed on the road at lower temperatures than the conventional hot-mix method. By extending the construction season, warm mix asphalt allows projects to be completed faster.
Prefabricated bridge elements	An old bridge can be demolished while the new bridge elements are built at the same time off-site, then brought to the project location ready to erect.
Integrated bridge support technology	Use of alternating layers of fill material and fabric reinforcement sheets to provide support for small bridges and reduce construction time.
Pavement Safety Edge	Saves lives by allowing drivers who stray off highways to return to the road safely. Instead of a vertical drop-off, the Safety Edge shapes the edge of the pavement to an angle that allows drivers to reenter the roadway safely.
Adaptive signal control technology	This technology adjusts the timing of red, yellow, and green lights to accommodate changing traffic patterns and ease traffic congestion.

Source: FHWA.

[a] A mitigation bank is a large-scale mitigation site approved through the federal Clean Water Act and other state programs for resource protection. In mitigation banking, infrastructure developers buy credits from the bank sponsor before any impacts occur. An in-lieu fee program conducts wetland, stream, or threatened or endangered species habitat restoration, creation, enhancement, or preservation. In-lieu fee programs must be administered by government entities or nonprofits. Users of the program pay the program sponsor for mitigation credits. See more detailed explanations of these programs in appendix II.

FHWA has developed performance measures for Every Day Counts that are linked to U.S. DOT performance measures, but it is too soon to determine the effect these initiatives have had on expediting highway projects, according to an FHWA official. FHWA and other highway project stakeholders developed one or two performance measures for each of the Every Day Counts initiatives. For example, FHWA established the following performance measure for the "Expanding Use of Programmatic Agreements" under Every Day Counts: "FHWA will expand, revise, or create 15 programmatic agreements at the state and regional scale by December 30, 2011." These performance measures support the overall goals of the Every Day Counts effort, which the FHWA administrator has stated are "to cut project delivery time in half and more quickly advance innovation into daily practice." The Every Day Counts performance measures and goals that FHWA established are linked to a U.S. DOT performance measure, an attribute of successful performance measures.[17] Specifically, since fiscal year 2010, U.S. DOT has had a performance measure to streamline environmental review with a target of 48 months to complete an EIS for major transportation projects.[18] U.S. DOT noted in its fiscal year 2011 performance report that Every Day Counts is an effort to help reduce project times. Every Day Counts includes 13 specific initiatives to streamline time frames for all four phases of highway projects, including the environmental review phase. While FHWA has collected data to address the U.S. DOT target noted above and data on the Every Day Counts initiatives, states have only had about 1 year to implement the Every Day Counts initiatives and, according to an FHWA official, it is too soon to tell if those initiatives have had a positive effect on expediting the completion of highway projects.

For fiscal year 2011, FHWA reported that the median time to complete EISs was 79 months (about 6.6 years). However, FHWA also reported that 16 of the EISs completed during fiscal year 2011 were started before August 2005, when SAFETEA-LU was enacted, and the median time to complete those EISs was 110 months (about 9 years). For the 7 remaining EISs completed during 2011, the median time to completion was 44 months (a little under 3.7 years). These data suggest that the full impact of the Every Day Counts initiatives on the time to complete EISs may not be discernable for several years, if ever, due to a number of reasons. Such reasons could include (1) the lengthy time frames needed to adopt complex initiatives such as linking planning and environmental review or compensatory mitigation,[19] (2) the possibility of EISs being completed which had started before SAFETEA-LU was enacted or Every Day Counts began, or (3) the impact of the SAFETEA-LU provisions discussed above that were also meant to expedite highway projects.

FHWA's efforts to share promising practices depend on the willingness of state DOTs to adopt them. Each state has identified Every Day Counts initiatives to use, according to an FHWA report. An FHWA headquarters official provided the following examples of Every Day Counts initiatives that have achieved wide acceptance:

- States have, since the introduction of Every Day Counts, generated 56 new programmatic agreements, far more than the Every Day Counts goal of 15 programmatic agreements by December 2011.
- Nearly all states are using warm mix asphalt. Interest in the product increased after FHWA promoted it as an Every Day Counts initiative.
- Over 40 states are using prefabricated bridge elements. They report working on 663 bridges, far more than the Every Day Counts goal of designing or building 100 bridges by December 2012.

CONCLUDING OBSERVATIONS

Completing major highway projects involves a complex process that depends on a wide range of stakeholders conducting many tasks. The long time frames to complete highway projects are often caused by factors outside the control of state DOTs, such as a lack of available funds, changes in a state's transportation priorities, or litigation. These factors can be project specific and may not be controllable by legislation, or by federal or state initiatives.

The SAFETEA-LU provisions meant to help expedite highway projects are generally viewed by state DOTs as having the potential to save time. However, given that state DOTs noted in our survey that there are other solutions outside of the SAFETEA-LU provisions that better serve their needs and are within their authority to implement, it is unlikely that state DOTs will greatly increase their participation in some of the SAFETEA-LU provisions we analyzed, particularly those that delegate environmental review decision-making authority from FHWA to state DOTs and require the state to accept federal court jurisdiction for such decisions. Regardless, keeping these provisions in law would continue to give state DOTs the ability to pursue these provisions should they later choose to do so.

FHWA's Every Day Counts effort offers a structured approach to collecting and sharing information with state DOTs to help expedite highway projects. FHWA's continued efforts to (1) track the progress of Every Day Counts using the performance measures it developed for each initiative and (2) use Every Day Counts as a way to keep introducing new initiatives for trial and adoption by state DOTs can help to ensure that promising practices are developed and shared among states. Additionally, use of the Every Day Counts effort could help U.S. DOT as it attempts to meet its performance measure to streamline environmental review.

AGENCY COMMENTS

We provided U.S. DOT with a draft of this report for review and comment. U.S. DOT provided technical comments, which we incorporated as appropriate.

David J. Wise
Director, Physical Infrastructure Issues

APPENDIX I. OBJECTIVES, SCOPE AND METHODOLOGY

Our work for this report was focused on federal-aid highway projects and efforts to expedite those projects. In particular, this report addresses the following questions: (1) What is the process for planning, designing, and constructing federally funded highway projects, and what factors could affect project time frames? (2) What are state departments of transportation (DOT) views on the benefits and challenges of implementing initiatives to expedite highway projects established by the Safe, Accountable, Flexible, Efficient Transportation Equity Act: A Legacy for Users (SAFETEA-LU)? (3) What practices have state DOTs and the Federal Highway Administration (FHWA) implemented to expedite highway projects?

To describe the process for completing highway projects, as well as the factors that could affect project time frames, we reviewed and analyzed relevant legislation—particularly SAFETEA-LU—regulations, congressional hearing statements, and relevant reports and other publications. We also conducted a number of interviews in six states to collect information on practices involved in completing highway projects, as well as factors that could affect project time frames (see table 4). We chose these six states using several criteria, including participation in the environmental review delegation programs established under SAFETEA-LU (i.e., the Categorical Exclusion Approval Authority and the NEPA Approval Authority provisions), number of active and inactive environmental impact statements, amount of federal highway funding received in fiscal year 2011, and geographic location within the United States. The six states we chose to interview were Alaska, California, Missouri, North Carolina, Pennsylvania, and Utah. In each state, we interviewed the FHWA division office and the state DOT. To obtain more detailed information on processes to complete highway projects—but to minimize the burden on interviewees and in the interest of time—we interviewed regional offices of key resource agencies in two of the six states we selected: California and North Carolina.[20] We chose these states based on geographic diversity, as well as previously conducted fieldwork. We selected six resource agencies to interview, as they were often cited in our preliminary review of reports, publications, and other documents, as well as in early interviews with state DOTs. The resource agencies we interviewed were: U.S. Army Corps of Engineers, U.S. Environmental Protection Agency, U.S. Fish and Wildlife Service, U.S. Forest Service, National Marine Fisheries Service, and state historic preservation offices. There are potentially other federal—as well as state, local, and tribal—agencies that could have been interviewed, but we chose to limit our scope to these six federal agencies. These interviews are not generalizable to all states. Furthermore, for this report, we focused only on federal-aid highways and not other types of highways.

Table 4. Highway Project Stakeholders Interviewed

Agency	Headquarters office	Alaska	California	Missouri	North Carolina	Pennsylvania	Utah
FHWA	✓	✓	✓	✓	✓	✓	✓
State DOT		✓	✓	✓	✓	✓	✓
Advisory Council on Historic Preservation/State Historic Preservation Office	✓		✓		✓		
U.S. Army Corps of Engineers	✓		✓		✓		
U.S. Environmental Protection Agency	✓		✓		✓		
U.S. Fish and Wildlife Service	✓		✓		✓		
U.S. Forest Service	✓		✓		✓		
National Marine Fisheries Service	✓		✓		✓		

Source: GAO.

To identify state DOT perspectives on the benefits and challenges associated with implementing provisions meant to help expedite highway projects established in SAFETEA-LU, we (1) reviewed information obtained in the above mentioned interviews and (2) conducted a survey of state DOTs. To conduct this survey, we identified key provisions within SAFETEA-LU that we felt were meant to expedite highway projects based on our review of the legislation, analysis of relevant reports, and interviews with highway project stakeholders. We then developed a draft survey to gather state DOTs' perspectives on the benefits and challenges associated with these SAFETEA-LU provisions. We selected five states in which to conduct pretests: California, Iowa, Pennsylvania, Utah, and Washington. In each pretest, we provided a state DOT official with a copy of our draft survey, asked them to complete it, and then contacted them after 1 hour to discuss the clarity of each question. Through this method, we were able to refine the questions and closed-ended responses in our survey. After the five pretests were completed, we provided a draft copy of the survey to FHWA and AASHTO for their review and comment. Both organizations provided technical comments that we incorporated, as appropriate. Using our professional judgment based on early interviews with highway project stakeholders and our pretests, we determined that the survey should be sent to environmental officials at the state DOTs. However, because the survey considered other aspects of highway projects—for example planning, right-of-way acquisition, and construction—language was included in our transmittal e-mails and in the introduction of the survey to indicate that the state DOT official receiving our survey should consult with his or her colleagues when completing it. We felt that it would be far more cumbersome for respondents, and potentially less reliable, if we were to develop and transmit separate surveys for each highway project phase—that is, individual surveys that covered planning, preliminary design and environmental review, final design and right-of-way acquisition, and construction. Thus, one survey was developed and respondents were asked to share it and consult with colleagues when providing responses. We used lists of environmental officials at the state DOTs that were compiled by AASHTO and the Transportation Research Board to determine the relevant survey respondents. The full universe for this survey was 52 state DOTs: all states, the District of Columbia, and Puerto Rico. We took steps, such as sending early notification e-mails, to help ensure that the list of respondents we created was accurate. We launched our survey on November 30, 2011. We sent e-mail reminders and telephoned survey respondents who had not completed the survey, urging them to do so as soon as possible. We eventually received responses from all 52 state DOTs. We reviewed these responses for inaccuracies or omissions, analyzed the data, and have presented the key findings in this report. The survey and its responses—with the exception of open-ended responses or other identifying information—is reproduced in our e-supplement for this report: see GAO-12-637SP .

While all state DOTs were included in our survey and, therefore, our data are not subject to sampling errors, the practical difficulties of conducting any survey may introduce nonsampling errors. For example, differences in how a particular question is interpreted, the sources of information available to respondents, or the types of people who do not respond to a question can introduce errors into the survey results. We included steps in both the data collection and data analysis stages to minimize such nonsampling errors. We collaborated with GAO survey specialists to design draft questionnaires and, as previously noted, versions of the questionnaire were pretested, revised, and sent to FHWA and AASHTO for review and comment. We examined the survey results and performed computer analyses to identify inconsistencies and other indications of error and addressed such issues, where possible. A second, independent analyst checked the accuracy of all computer analyses to minimize the likelihood of errors in data processing. In addition, GAO analysts answered respondent questions and resolved difficulties that respondents had in answering our questions.

To describe the practices state DOTs have implemented on their own to help expedite highway projects, we included a series of questions in our survey of state DOTs asking respondents to identify practices they have implemented in each highway project phase: planning, preliminary design and environmental review, final design and right-of-way acquisition, and construction. State DOTs provided these responses in open-ended questions, which we analyzed. The practices we identified in this report were those that were cited most frequently by survey respondents. To collect additional information on efforts both state DOTs and FHWA have implemented to help expedite highway projects, we reviewed and analyzed information obtained during the interviews with FHWA (both headquarters and division offices), state DOTs, and resource agencies.

APPENDIX II. INITIATIVES IN THE FEDERAL HIGHWAY ADMINISTRATION'S EVERY DAY COUNTS EFFORT

The Federal Highway Administration (FHWA) is sharing information on methods to expedite highway projects with state departments of transportation (DOT) through an effort called Every Day Counts. This effort's goals are to shorten project time frames and accelerate technology and innovation by convincing states to adopt proven, rapidly deployable innovations. Under Every Day Counts, FHWA urged state DOTs to consider use of 15 specific initiatives. Every Day Counts promoted the following 10 initiatives to shorten project time frames:

- *Linking planning and environmental review.* This initiative promotes use of planning documents and decisions from the project planning process in the environmental review process. It takes environmental, community, and economic information collected early in the planning stage and carries it through project development, design, and construction. This can lead to decision making that minimizes duplication of effort, promotes environmental stewardship, and reduces delays in project implementation.
- *Early consultation with FHWA environmental attorneys.* Decisions made early in planning and project development are often the root causes of problems identified later in the environmental review process when National Environmental Policy Act of 1969 (NEPA) and Section 4(f) documents undergo legal scrutiny. Consulting with FHWA environmental attorneys at early decision points can help decision makers avoid problems later, saving time and costs.
- *Expanded use of programmatic agreements.* A programmatic agreement is a document that formally spells out the terms of an agreement between a state DOT and other state and/or federal agencies. A programmatic agreement establishes a process for consultation, review, and compliance with one or more federal laws. According to FHWA, programmatic agreements have been effective in producing time savings for completing highway projects.
- *Compensatory mitigation.* The federal Clean Water Act requires compensatory mitigation for projects that cause unavoidable impacts to streams, wetlands, and other waters of the United States. Mitigation for federally protected species may also be required through the

Endangered Species Act. Some state laws and regulations also require compensatory mitigation. The permitting process under Section 404 of the Clean Water Act constitutes a major component of the project development and completion process. This initiative proposes expanded use of in-lieu fees and mitigation banking in order to save time and expedite highway projects. See table 5 for information on these compensatory mitigation approaches. Some states have never used mitigation banks or in-lieu fee programs, while others use them for the majority of their mitigation needs.

- *Clarifying scope of preliminary design.* Some consider preliminary design to involve only the activities needed to make a NEPA determination; they view everything else as final design activities. This cautious approach delays highway projects because it postpones essential planning until it is too late to be effective. States have the flexibility to pursue many design activities not required for a NEPA determination under preliminary design. When performed concurrently with the NEPA process, these activities can expedite projects without affecting eligibility for federal aid. For example, states can perform soil borings, preliminary traffic control plans, and grading plans.
- *Allowable right-of-way acquisition streamlining.* Before building a highway project, land and property must be acquired by federal, state, and local agencies through right-of-way practices and procedures. Instead of sequentially, these agencies may move elements of a project through the right-of-way process concurrently. This can significantly shorten the highway project development process. To save time, agencies can use these process flexibilities including
  - appraisal waiver valuations,
  - incentive payments to advance acquisition and relocation, or
  - appraisals and negotiations of property acquisition (up to $10,000) by the same individual.
- *Effective coordination for utility relocation.* Approximately half of all highway and bridge projects eligible for federal funding require the relocation or adjustments to accommodate utilities. Gas lines, water lines, waste plumbing, electrical wires, telephone lines, and other wiring are often affected by highway and bridge projects. Flexibilities in place under federal law and regulations foster effective utility coordination during project development by

- determining the best strategy for physically relocating utilities when needed,
- documenting the terms and considerations for accomplishing utility relocation activities, and
- financing the work in an effective and timely manner.

- *FHWA technical assistance for stalled projects*. This initiative focuses on new projects wherein problems are anticipated with conducting an effective project development process, or for "ongoing EISs" where 60 months or more have elapsed since the publishing of the project's notice of intent and no record of decision has been issued. FHWA technical assistance teams will resolve many issues that would otherwise hold up the NEPA review or otherwise delay a project's progress. FHWA subject matter experts can help resolve resource-specific issues concerning wetlands, endangered species, and cultural resources.
- *Use of design-build contracts*. Traditionally, a project is designed, put out for bid to construction firms, then built by the winning bidder (design-bid-build). As discussed above, design-build is an alternate contracting method in which the design and construction phases are combined into one contract, eliminating the separate bid phase and allowing certain aspects of design and construction to take place at the same time. This can provide significant time savings compared with the design-bid-build approach, where the design and construction phases must take place in sequence.
- *Use of construction manager/general contractor contracts*. In a construction manager/general contractor project, typically, the owners of a project are able to hire a general contractor early in the design phase so that the state may benefit from the contractor's constructability input as the design develops. This contract type allows state DOTs to remain active in the design process while assigning risks to the parties most able to mitigate them. This can save time because a number of activities can be undertaken concurrently. FHWA allows this type of contract only on a trial basis because approval is necessary for any nontraditional construction contracting technique that deviates from the competitive bidding provisions in Section 112 of Title 23 of the U.S. Code.

Table 5. Compensatory Mitigation Approaches Promoted by FHWA's Every Day Counts

Mitigation bank	In-lieu fee program
A mitigation bank is a large-scale mitigation site approved through the federal Clean Water Act and other state programs for resource protection. Each mitigation bank has a formal agreement between the bank sponsor and the Federal regulatory agencies. The agreement details the number of credits the bank can generate; the types of habitat the mitigation bank intends to create, restore, or enhance; and explains the long term management mechanism that will be utilized to ensure the site is protected in perpetuity. In mitigation banking, infrastructure developers buy credits from a bank sponsor before any impacts occur.	An in-lieu fee program conducts wetland, stream, or threatened or endangered species habitat restoration, creation, enhancement, or preservation. In-lieu fee programs may perform various environmental enhancement activities throughout a watershed rather than at one particular site. In-lieu fee programs establish a similar agreement to a mitigation bank agreement, but the sites are not always completely constructed prior to the environmental impacts taking place. Once enough money is received by the program, it implements the project in that watershed. Federal regulations require that in-lieu fee programs be administered by government entities or nonprofits. Users of the program pay a predetermined amount per mitigation credit to a program sponsor, who performs the actual mitigation.

Source: FHWA.

FHWA promoted five tools to accelerate use of innovative technology, including three that can shorten the time needed to complete highway projects:[21]

- *Warm mix asphalt paving.* Warm mix asphalt is the generic term for technologies that allow asphalt to be produced and then placed on the road at lower temperatures than the conventional hot-mix method. In most cases, the lower temperatures result in significant cost savings and reduce greenhouse gas emissions because less fuel is required. Warm mix asphalt also has the potential to extend the construction season, allowing projects to be completed faster.
- *Prefabricated bridge elements.* Use of prefabricated bridge elements means that many time-consuming construction tasks no longer need to be done sequentially in work zones. An old bridge can be demolished while the new bridge elements are built at the same time off-site, then

brought to the project location ready to erect. Because the bridge elements are usually fabricated under controlled climate conditions, weather has less impact on the quality, safety, and duration of the project. The use of prefabricated bridge elements also offers cost savings. The ability to rapidly install prefabricated bridge elements on-site can reduce the environmental impact of bridge construction in environmentally sensitive areas. See a photograph of prefabricated bridge elements being assembled in figure 3.

- *Integrated bridge support technology.* Instead of conventional bridge support technology, an innovative bridge system technology uses alternating layers of compacted granular fill material and fabric reinforcement sheets to provide support for the bridge (see figure 4). The technology, known as geosynthetic reinforced soil technology, offers advantages in the construction of small bridges, including the following:
  - reduced construction time and cost,
  - ease of construction with common equipment and materials,
  - ease of maintenance, and
  - flexible design that is easily modified in the field for unforeseen site conditions.

Source: Jerome O'Connor, P.E., BridgeComposites, LLC.

Figure 3. Prefabricated Bridge Elements at Construction Site.

Source: FHWA.

Figure 4. Geosynthetic Reinforced Soil Technology at a Bridge Construction Site.

- *Safety Edge*. The Safety Edge, while not an initiative designed to save time in completing a highway project, is a simple but effective solution that can help save lives by allowing drivers who stray off highways to return to the road safely. Instead of a vertical drop-off, the Safety Edge shapes the edge of the pavement to 30 degrees—the optimal angle to allow drivers to reenter the roadway safely. FHWA's goal is to accelerate the use of the Safety Edge technology, working with states to develop specifications and adopt this pavement edge treatment as a standard practice on all new and resurfacing pavement projects.
- *Adaptive signal control technology*. Poor traffic signal timing contributes to traffic congestion and delay. Conventional signal systems use preprogrammed, daily signal timing schedules. Adaptive signal control technology adjusts the timing of red, yellow and green lights to accommodate changing traffic patterns and ease traffic congestion. Though not designed to save time in completing a highway project, the main benefits of adaptive signal control technology over conventional signal systems are that it can
  - continuously distribute green light time equitably for all traffic movements,

- improve travel time reliability by progressively moving vehicles through green lights,
- reduce congestion by creating smoother flow, and
- prolong the effectiveness of traffic signal timing.

End Notes

[1] For the 30 projects in which FHWA was the lead agency and signed the EIS in fiscal year 2010, the average amount of time from signing the notice of intent to signing the record of decision was 69 months.

[2] An additional provision that we reviewed—SAFETEA-LU Section 6003 "State assumption of responsibilities for certain programs and projects," codified at 23 U.S.C. § 325—has not yet been implemented. Numerical results from our survey on all the SAFETEA-LU provisions we studied, as well as more information on how we conducted our survey can be found in appendix I or in a separate e-supplement, GAO-12-637SP .

[3] As previously noted, FHWA estimates that based on its 2009 data approximately 1 percent of all federal-aid highway projects in the United States were processed with an EIS. The vast majority of highway projects—96 percent—were processed as CEs under NEPA and generally did not require as many tasks to complete.

[4] Specifically, 27 of 51 states (53 percent) identified funding as a "very great" or "substantial" challenge during the planning phase; 27 of 52 states (52 percent) identified funding as a "very great" or "substantial" challenge during the preliminary design and environmental review phase; 20 of 51 states (39 percent) identified funding as a "very great" or "substantial" challenge during the final design and right-of-way acquisition phase; and 17 of 50 states (34 percent) identified funding as a "very great" or "substantial" challenge during the construction phase.

[5] Se. Alaska Conservation Council v. Fed. Highway Admin., 649 F.3d 1050 (9th Cir. 2011).

[6] In our survey, we asked states to report the extent to which their agency agreed or disagreed that benefits, including time savings, could be realized from each of the SAFETEA-LU provisions we studied. For reporting purposes, we have combined responses of "strongly agree" and "agree," as well as "strongly disagree" and "disagree." Not all states responded to all questions asked in the survey.

[7] With respect to historic sites, a DOT may make a finding of de minimis impact if, among other things, it receives written concurrence from the applicable state historic preservation officer or tribal historic preservation officer. With respect to parks, recreation areas, or wildlife or waterfowl refuges, the DOT may make a finding of de minimis impact if, among other things, it receives written concurrence from the officials with jurisdiction over the park, recreation area, or wildlife or waterfowl refuge.

[8] 23 U.S.C. § 138(a) and (b).

[9] We asked states the extent to which they agreed or disagreed that the following two issues are challenges that could be faced: (1) programmatic agreements could serve their agency better than this initiative and (2) state or agency policy discourages the use of this initiative. Only 4 out of 37 states agreed or strongly agreed that the first issue could be a challenge, and 8 out of 34 agreed or strongly agreed that the second could be a challenge.

[10] We asked states to provide a written response to the following question: "What challenges, if any, could be faced from the statute of limitations provision established in SAFETEA-LU Section 6002?" Twenty-five states provided some form of a written response, with one state noting two challenges. Of those, 8 states provided a response indicating that a shorter statute of limitations could encourage others to file a lawsuit against a project; 10 provided a

response indicating that they did not have a comment or that the provision was beneficial; and 8 states' responses indicated a unique challenge.

[11] More specifically, 23 U.S.C. § 326(c)(3) states: "In a memorandum of understanding, the State shall consent to accept the jurisdiction of the Federal courts for the compliance, discharge, and enforcement of any responsibility of the Secretary [of Transportation] that the State assumes."

[12] Pursuant to SAFETEA-LU, the DOT Secretary may not assign responsibility for any conformity determination required under section 176 of the Clean Air Act or any responsibility imposed on the Secretary related to transportation planning as established in 23 U.S.C. §§ 134, 135. See 23 U.S.C. § 327.

[13] Of the states responding, 6 out of 35 agreed that staffing or personnel savings could be realized, and 7 out of 35 agreed that an increased number of projects could be completed from using this provision.

[14] For each of the four phases of highway projects, we asked states the extent to which they agreed or disagreed that time savings, staffing/personnel savings, and increased number of projects completed are potential benefits of initiatives they have undertaken. The numbers of states that agreed or strongly agreed that time savings are a potential benefit are: planning phase—18 out of 21 states; preliminary design and environmental review phase—36 out of 39 states; final design and right-of-way acquisition phase—15 out of 15 states; and construction phase—18 out of 18 states.

[15] The construction manager/general contractor method allows a state to hire a contractor early in the design process in order to benefit from the contractor's constructability input as the design develops. The method can save time, compared with the traditional method, because construction can start before designs are complete. Under the traditional method, design and construction phases must take place in sequence.

[16] *Cost plus time bidding* is a contracting procedure that selects the low bidder based on a monetary combination of the contract bid items and the time needed to complete the project. *Lane rental* is a contracting technique by which a contractor is charged a fee for occupying existing highway lanes or shoulders during construction. The fee motivates the contractor to minimize the time that a lane or a shoulder is out of service, minimizing traffic delay to highway users. *Contractor completion incentives and disincentives* are contract provisions that pay the contractor a certain amount of money for each day a highway project is completed ahead of schedule and charge money for each day the contractor overruns scheduled time.

[17] Past GAO work has identified nine attributes of successful performance measures, including linkage, meaning that performance goals and measures at the operational level should align with an agency's goals and mission. See GAO, *Tax Administration: IRS Needs to Further Refine Its Tax Filing Season Performance Measures,* GAO-03-143 (Washington. D.C: Nov. 22, 2002).

[18] The time to complete an EIS is generally measured from the date the notice of intent is issued to the date the record of decision is issued.

[19] One state, cited by FHWA as a case study at linking planning and environmental review, took 21 months getting from the summit meeting that introduced the new integrated planning system concept to the memorandum of understanding signed by 24 agencies that agreed to support the new system. Another state, cited by FHWA as a case study at implementing compensatory mitigation, spent 18 months developing its memorandum of agreement to implement the mitigation concept developed at a workshop.

[20] In general, the Advisory Council on Historic Preservation—the federal resource agency responsible for historic preservation—delegates much of its authority under the National Historic Preservation Act to state historic preservation offices.

[21] In addition to the innovations that expedite projects, Every Day Counts included two innovations that serve other purposes—a beveled-edge pavement that improves highway safety and adaptive signal control technology that reduces travel times and traffic delays.

INDEX

A

B

C

D

E

F

G

H

I

Q

R

S

W